FUNGAL SPORES

Their Liberation and Dispersal

FUNGAL SPORES

Their Liberation and Dispersal

C. T. INGOLD

CLARENDON PRESS · OXFORD

1971

Oxford University Press, Ely House, London W. 1

GLASGOW NEW YORK TORONTO MELBOURNE WELLINGTON
CAPE TOWN SALISBURY IBADAN NAIROBI DAR ES SALAAM LUSAKA ADDIS ABABA
BOMBAY CALCUTTA MADRAS KARACHI LAHORE DACCA
KUALA LUMPUR SINGAPORE HONG KONG TOKYO

PRINTED IN GREAT BRITAIN
AT THE PITMAN PRESS, BATH

PREFACE

INSTEAD of providing second editions of *Dispersal in Fungi* and *Spore Liberation*, I decided to unite the two into a single work, omitting, however, the discussion of bryophytes that occurs at the end of the latter. Initially an exercise involving scissors and paste was envisaged, but it soon appeared that this was not suitable, and a complete re-writing had to be undertaken. This is largely because so much has been done in the general field of dispersal in fungi during the past decade. Further, although many illustrations from the earlier books are included in the present one, it was necessary to provide over 80 new ones.

Although a large proportion of the illustrations are my own, many are from other authors and I am grateful to them for allowing me to reproduce their figures. Further, I have to thank the Director of the Royal Botanic Gardens, Kew for permission to use two previously unpublished drawings by the late Professor A. H. R. Buller, that occur in the typescript of an unpublished volume of his *Researches* bequeathed to Kew. My thanks are also due to the Hafner Publishing Company who have given permission to reproduce certain figures from Buller's published volumes. I am also indebted to Dr. A. D. Greenwood for allowing me to reproduce stills from his film of *Saprolegnia*.

I believe that students of fungi will always be interested in questions of spore dispersal, but we are now in an age when the study of mechanisms is nearing its end, and emphasis will probably be on the extension of knowledge by the quantitative evaluation of dispersal in the overall ecological picture of fungi in field situations.

Birkbeck College   C. T. INGOLD
London, 1970.

CONTENTS

I. INTRODUCTION

THIS book is concerned with the liberation and dispersal of fungal spores, but before turning our attention to the processes involved it is important to consider the actual necessity for dispersal.

Dispersal seems to be a problem for all kinds of organism. Each species of plant or animal occurs in a circumscribed geographical area; it has a fairly definite range. This may be of small extent. Indeed a species may be limited to one isolated little island or to a single mountain top. On the other hand it may range over almost the whole of the habitable world. Even where the range is now great, the species almost certainly started in one spot. It has been argued that the widely-ranging species tend to be the older, while comparable species that at present occupy small areas tend to be young beginners. Extension of geographical range is a feature of the history of each species and for this extension some mechanism of dispersal is a necessity.

Most animals exploit their immediate territory and extend their range by their own movements, but the fixed organism, whether an animal such as a sponge or coral, or a plant, must rely on detachable units for dispersal. Sponges and corals have their free-swimming larval stages, flowering plants their seeds, and fungi, ferns, mosses, liverworts, and seaweeds their microscopic spores.

The necessity for dispersal is not confined to the extension of geographical area. Individuals within the general range of each species are usually limited to certain ecological niches and if these are scattered the dispersal mechanism must be adequate to provide propagules in the right places when and where opportunities offer.

Again, dispersal has a genetic importance. Each species at any time is more or less in equilibrium with its environment, both physical and biotic. But this environment may change over the years and a species, if it is to survive, must be capable of adjusting itself to these altering conditions. Here the degree of genetic variability available for selection may be of great importance. Dispersal may be of significance in giving the opportunity for new variability, when it arises in a species at a certain point in its range, to spread among the whole population. This spread of new variability may be achieved by the dispersal of reproductive units, such as seeds and spores, capable of giving rise to new individuals, but it may also be achieved by the spread of pollen grains in higher plants. A similar example from the fungi is the dispersal of pycniospores of rusts which can, like pollen grains, transport genes, but cannot grow directly into new individuals. It should be noted, however, that although efficient dispersal may spread new genes among a population and so increase the evolutionary plasticity of a species, it tends to operate against actual speciation, since effective dispersal breaks down isolation on which species differentiation so largely depends.

To sum up, it may be said that dispersal is of significance for the maintenance of the population within its existing range, for the extension of the range of a species, and for its genetical development.

Fungi reproduce and spread mainly by spores. These, as in other cryptogams, are microscopic units mostly unicellular although not infrequently multicellular and containing some food reserve, usually oil or glycogen. Many fungal spores are meiospores, as in bryophytes and pteridophytes, with a meiosis involved in their formation. Ascospores, basidiospores, and the spores in the sporangia of Mycetozoa are of this nature. However, spore production may be quite unrelated to meiosis. This is true, for example, of the great range of conidial forms classified in Fungi Imperfecti, of the conidial stages of Ascomycetes, of the urediospores of Uredinales (rusts), and of the sporangiospores of Mucorales.

Nearly all spores are essentially dispersive units. Some, however, are merely resting structures that can tide the fungus over an unfavourable period such as the cold of winter or prolonged drought. To this category belong most rust teliospores, although some are also dispersive, and the oospores of Phycomycetes. Again, zygospores of Mucorales are resting, rather than dispersal, spores, but the part they play in the general biology of these fungi is far from clear.

The great majority of spores have firm cell walls, but the zoospore of water-moulds is naked, although having come to rest it then secretes a wall prior to germination.

Although essentially microscopic, spores of fungi vary greatly in size and shape (Fig. 1.1). Most are, however, spherical or ovoid with a diameter in the range 5–50 μm, but the ascospores of some lichens are nearly visible with the unaided eye. For example, the two-celled ascospore of *Varicellaria microsticta* may be as large as 350 × 115 μm, nearly as big as the smallest orchid seeds. Many spores are long and thread-like. Thus the septate ascospores of *Cordyceps militaris* may be 500 μm long but only 2 μm wide. In some fungi the spore is a branched structure, a feature particularly characteristic of aquatic Hyphomycetes. Although single spores are microscopic, in the mass they may become conspicuous as in the spore print of a toadstool, or in the smoke-like cloud rising from a puffing *Peziza*, or from a ripe puff-ball (*Lycoperdon* sp.) bombarded by falling raindrops.

Spores vary in colour. Many are transparent and colourless, appearing white in the mass; but they may be yellow, pink, purple, brown, or black,

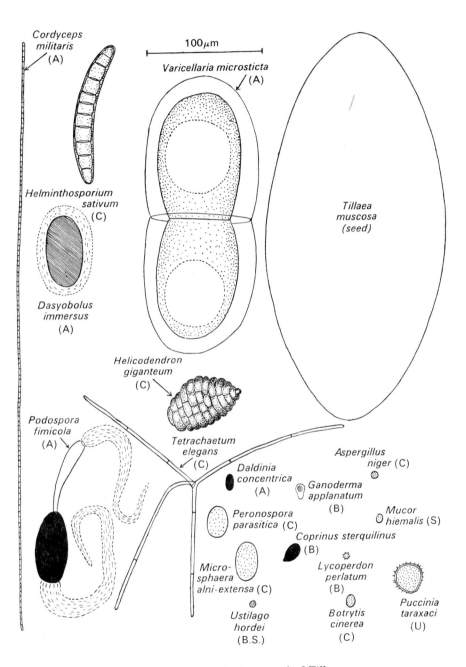

Fig. 1.1. Size of spores. Outline of minute seed of *Tillaea muscosa*, seven exceptionally large fungal spores, and eleven spores of more usual size and form. All drawn to same scale. Type of spore indicated in brackets with each species. (A) ascospore; (B) basidiospore; (B.S.) brand spore; (C) conidium; (U) urediospore; (S) sporangiospore.

and indeed spore colour is an important taxonomic character, particularly in the Agaricales. The colour of spores is due largely to pigmentation in the spore wall, although yellow-orange carotinoid pigments dissolved in oil drops in the cytoplasm may also contribute, particularly in the urediospores of rusts.

Apart from pigmentation, the wall of the spore may vary considerably. It can be thin or thick and either smooth or variously ornamented.

An outstanding feature of most fungi is the enormous spore production. However, on average not more than one spore from each individual succeeds in its reproductive function, since each species is more or less in equilibrium and its numbers, though they may fluctuate from year to year, usually show no steady increase.

Many estimates of spore output have been made, but only a few examples will be given to illustrate its gigantic scale in a wide range of fungi. A big specimen of the giant puff-ball (*Calvatia gigantea*) has been estimated to contain 7 000 000 000 000 spores. The large bracket fungus *Ganoderma applanatum* may discharge 30 000 000 000 spores a day, apparently maintaining this output for the whole five months (May to September) of its annual spore-fall period. The small apothecium of *Sclerotinia sclerotiorum* has been shown to produce 30 000 000 ascospores. A perithecial stroma of the flask-fungus *Daldinia concentrica* may discharge over 100 000 000 spores a day. In the stinking smut of wheat (*Tilletia caries*) a single diseased grain may contain 12 000 000 brand spores. A colony of blue mould (*Penicillium* sp.) 2·5 cm in diameter may bear 400 000 000 conidia.

An important attribute of a spore from the point of view of dispersal is its retention of the power to germinate. The fact that it may be transported a long distance is of no significance if at the end of its journey it is incapable of growth. The bearing of this on effective dispersal is well illustrated by rusts. On the whole urediospores are very resistant, and being able to survive for a long time in the air, in spite of both dessication and intense insolation, can carry rust infection in a single step to a distance of hundreds of miles. In contrast the smaller, thin-walled basidiospores (sporidia) are short-lived and are seldom capable of causing infection at a distance of more than a few miles from diseased plants. However, all too little is known about the retention of viability in spores during the course of their dispersal.

The great majority of fungi are spread by spores, but some employ other means of dispersal. In the minute agaric *Omphalia flavida*, for example, which causes a leaf-spot of coffee and many other plants in the New World, although fruit-bodies are formed liberating basidiospores, reproduction also occurs by macroscopic gemmae (Fig. 1.2). Each of these appears to be homologous with a pileus, but much smaller. When ripe it becomes loosened from its stipe, is readily blown away, and, alighting on a suitable substratum, gives rise to a new mycelium (Buller 1934). In *Sclerotium coffeicola*, another coffee parasite, reproduction and dispersal are entirely by gemmae comparable with those of *O. flavida* (Buller 1934). Again, the sclerotium of ergot (*Claviceps purpurea*), like the host caryopsis that it replaces, becomes detached and may play a part in dispersal, although the spread of this fungus is mainly brought about by wind-borne ascospores and insect-borne conidia.

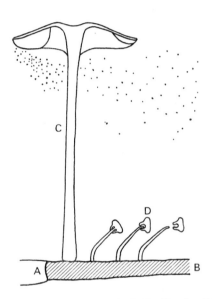

FIG. 1.2. *Omphalia flavida* parasitizing leaf of coffee. Leaf tissue in vertical section: (A) living; (B) dead; (C) sporophore liberating basidiospores; (D) gemmifers each bearing a terminal gemma. C× 5, but D× 10. After figures by Buller (1934).

The commonest type of vegetative reproduction in fungi is found in lichens, which are frequently dispersed by soredia. In many lichens (e.g. *Cladonia* spp.) there occurs on the surface of the thallus a dry, greyish powder each grain of which is a single soredium consisting of a tangled ball of fungal hyphae wrapped around a few algal cells. This powder is readily blown away by the wind. Many species rely for reproduction and dispersal entirely on soredia.

In the air spora (p. 175) other reproductive units occur which are not spores. Pady and Gregory (1963) report the normal occurrence in the air of hyphal fragments many of which are capable of growth. Further, somewhat as bracken spreads over a hill-side by growth of its underground rhizome, a few fungi are locally dispersed by growth of an underground rhizomorph, for example in *Armillaria mellea*.

The dispersal story of any fungus is usually divisible into a number of episodes and for most species, as with aircraft, there are three: take-off, the actual flight (transport), and landing. However, not all fungi conform neatly with this pattern. Thus in the splash dispersal of slime-spore species the three episodes are part of what is really a single process. Again, in such a coprophilous fungus as *Coprinus*, following the initial take-off, flight, and landing of the spores on the grass, there are further stages involving passage through an animal.

Since the great majority of fungi are dispersed through the air the behaviour of spores when suspended in air is of special interest. A significant characteristic is the rate of fall in still air, although the motion of the air itself, particularly its degree of turbulence, may have over-riding importance. This latter question will be discussed in detail later.

For a spherical spore, fall in still air is governed by Stokes' law which can be expressed by the following equation:

$$V = \frac{2}{9} \cdot \frac{\sigma - \rho}{\mu} \, gr^2,$$

where V is steady terminal velocity in cm/s,
    σ is density of spore,
    ρ is density of air,
    g is acceleration due to gravity (981 cm/s),
    μ is viscosity of air ($1 \cdot 8 \times 10^{-4}$ g/cm/s at 18°C), and
    r is radius of spore.

Since the density of most spores is approximately $1 \cdot 0$ and that of the air is so small that it can be ignored, this can be simplified to

$$V = \frac{2gr^2}{9\mu}.$$

Thus the rate of fall of a spherical spore is proportional to the *square* of its radius. There is reasonable agreement between theory and observed values.

Rates of fall actually measured vary between $0 \cdot 05$ cm/s for the spore of *Lycoperdon pyriforme* (4 μm diam.) to $2 \cdot 0$–$2 \cdot 8$ cm/s for the very large spores of *Helminthosporium sativum* (80×15 μm). Where the spore is elongated its terminal velocity is reduced as compared with a sphere of the same volume. Thus an elongated ellipsoid with the long axis four times that of the short axis is retarded by a factor of $1 : 1 \cdot 28$ (Chamberlain 1967).

The Stokes equation, which describes the steady rate of fall V of a particle in still air, is important in another connection. Many fungal spores are violently discharged into the air and V, as calculated above, also enters into the equation denoting the horizontal distance D to which a spore, with a given initial velocity H, is shot into the air. The equation, given by Buller (1909), is

$$D = \frac{HV}{g}.†$$

This, however, does not take into consideration the aerodynamic drag effect on a particle moving through the air initially at a high speed. Figure 1.3 illustrates the position when correction is made for this factor (technically the Reynolds number of the particle). Where the lines are straight no correction is involved. The situation for the majority of fungal spore-projectiles, with the exception of the sporangia of *Pilobolus* and the glebal mass of *Sphaerobolus*, falls within the area of the figure. For most mycological purposes, the Reynolds number correction can safely be ignored.

The great majority of Fungi are terrestrial and this book is largely concerned with the transport of spores through the air, whether carried by wind, splash-borne, or moved by insects. A very much smaller number are, however, submerged aquatics, and dispersal of their spores through the water will be considered separately (Chapter 16).

In very many species the spores are initially launched into the air violently. Usually when this is so turgid living cells are involved. In some fungi, particularly Ascomycetes (Chapter 2), a water-squirting mechanism is the rule, while in others sudden rounding-off of turgid cells (Chapter 4) as with the aeciospores in the aecia of rusts, accounts for discharge. In the largest section of Basidiomycetes, Hymenomycetes, externally produced 'ballistospores' are discharged by a mechanism that still remains obscure (Chapter 5).

In a few fungi where the spores are actively discharged, the release mechanism is operated as a result of evaporation from cells (Chapter 7). This usually involves jerking spores off their conidiophores either by sudden twisting movements produced by rapid water loss or by the instantaneous separation of a gas phase in differentially thickened cells of the conidiophore distorted by drying. When the gas phase suddenly appears, the conidiophore returns immediately to its original shape.

Passive liberation by blow-off, splash-off, and shake-off is of importance in many fungi (Chapter 9). Further, although fungi are predominantly wind dispersed (Chapter 10) it is also necessary to consider dispersal by insects and other animals (Chapters 13 and 14). Again the spread of fungi with the reproductive units of higher plants has practical significance, so that some attention to seed-borne fungi is imperative (Chapter 15).

† Essentially the same equation has been developed by Chamberlain (1967).

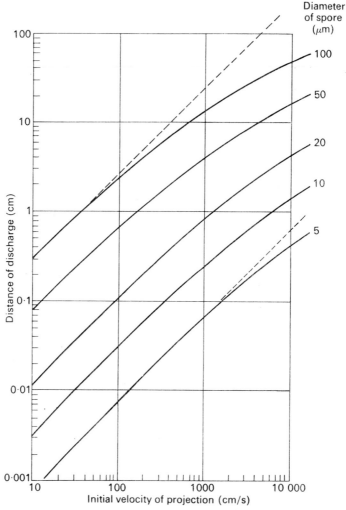

Fig. 1.3. Distance of discharge of unit density spheres (of five sizes) projected into still air plotted against initial velocity of discharge. Dashed lines
for spores of 5 μm and 100 μm diameter show what would be the graph if no
account had to be taken of aerodynamic drag. The initial velocity of all fungal
spores is probably less than 2000 cm/s. Based on diagram by May (1967).

Throughout, in discussing liberation and dispersal, particular attention
is given to the experimental approach and the physiological processes
involved. It should be said that, although an attempt has been made to
cover all the major aspects of the subject, it is inevitable that certain of
these that have been of special interest to the author should be given
rather greater emphasis than they deserve, while others have received a
more cursory treatment.

2. THE LIBERATION OF SPORES IN ASCOMYCETES

IN the great majority of Ascomycetes, the largest group of the Fungi, the ascus is a turgid cell that finally bursts in a regular manner violently liberating its contained ascospores. There are, however, many examples, widely scattered in any scheme of natural classification, in which the ascus is not explosive. Some would regard all such Ascomycetes as degenerate in this respect, but others would consider certain genera as primitive (e.g. *Endomyces*) while agreeing that certain other types with non-explosive asci (e.g. *Genea, Chaetomium, Ceratocystis*) are probably degenerate in their ascus behaviour.

The typical ascus, just before it bursts, is a turgid cell (Fig. 2.1). The thin, stretched elastic ascus-wall is lined by a very thin layer of enucleate protoplasm that surrounds the large central vacuole of cell-sap in which the ascospores are suspended towards the apex of the ascus. The assumption is that the lining layer of cytoplasm behaves as a more or less semi-permeable membrane. The sap of the vacuole probably has a high osmotic pressure, although precise determinations are largely lacking. An estimate for *Sordaria fimicola*, based on an analysis of the sap accompanying the discharged spores, is 10–30 atm (Ingold 1966). The only sugar present seems to be glucose, but in *Sordaria* this accounts for a small proportion of the total osmotic pressure. Salts appear to make a more important contribution. It has generally been thought that the later stages of maturation in the ascus involve the hydrolysis of glycogen to sugar giving a rapid build-up of osmotic pressure. In many Ascomycetes the young ascus stained with iodine gives a chestnut-brown reaction indicative of glycogen and this reaction is no longer apparent at an older stage. However, it seems unlikely, in *Sordaria* at least, that hydrolysis of glycogen to glucose is a major factor in the development of the osmotic pressure of the ascus.

Although the hydrostatic pressure in most explosive asci is probably developed osmotically, in some the swelling of mucilage may be the essential mechanism. In species of the submerged aquatic genus *Loramyces* there is no central vacuole in the ascus and, apart from the spores themselves, it is their mucilaginous sheaths that occupy most of the interior (Fig. 2.1A). The pressure that finally leads to the bursting of the ascus with the discharge of the spores may well be generated by the swelling of this hydrophilic mucilage (Ingold 1968).

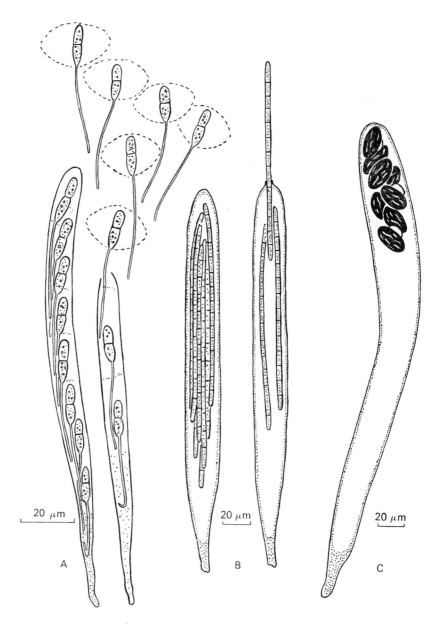

FIG. 2.1. Types of asci. (A) *Loramyces juncicola*: ascus before and during
spore liberation; clear areas in ascus represent mucilage and each liberated
spore has a mucilage sheath around its 'head'. (B) *Trichoglossum hirsutum*:
one ascus just before discharge; another from which five spores have already
successively escaped. (C) *Ascobolus stercorarius*: ascus ripe for discharge and
largely filled by vacuole of cell sap in which the spores are suspended.

The asci burst in a fairly definite way (Plate I): by a hinged lid, associated with a relatively large aperture, in most Pezizales (operculate Discomycetes); by an apical slit in *Sphaerotheca* and *Ascozonus*; by the complete separation of an apical cap in *Podospora* spp. and *Dasyobolus immersus*; or by the development of a minute pore in Helotiales (inoperculate Discomycetes), many Sphaeriales, and Clavicipitales.

The mechanism of dehiscence of inoperculate asci is far from clear. Perhaps a hydrolysis of material filling the pore is involved. Chadefaud (1969) and other French mycologists have devoted considerable study to the apical structure of these asci and have revealed complex systems, but nothing is known of the significance of these in relation to spore release.

A striking feature of explosive asci is the large range in size and form of the ascospores (Fig. 2.2). Many are unicellular and ovoid and spores of this shape may range from about 5 μm in length to the ascospore of the common lichen *Pertusaria pertusa* which is about 250 μm in length. Some are long and thread-like such as the ascospore of *Cordyceps militaris*, about 500 μm long but only 2 μm wide, which breaks up after discharge into over a hundred part-spores each 4×2 μm. A great many ascospores are more or less fusiform and septate (e.g. in *Leptosphaeria* spp.).

Associated to some extent with this great range of size is a considerable range in distance of spore discharge. This may be a millimetre or less (e.g. in *Cordyceps militaris*) but may reach 30–40 cm in *Dasyobolus immersus* and *Podospora fimicola*. In most Ascomycetes, however, the range is 0·5–2·0 cm. We shall return to this question later. By contrast with violently discharged ascospores, in basidiospores that are shot from their sterigmata and in other ballistospores, the range in size and form is strictly limited (Fig. 2.2) and associated with this the distance of ballistospore discharge is confined within the narrow limits of 0·005–0·1 cm.

In most Ascomycetes complex sporophores or fruit-bodies occur, and in the larger forms a single fruit-body may produce millions of asci. A number of rather distinct types of organization can be recognized in connection with the active release of ascospores.

In the *Discomycete* type the spore-producing surface (hymenium), consisting of elongated asci intermixed with parallel paraphyses, is more or less exposed, most often as a lining to a shallow cup-or saucer-shaped apothecium. The extensive hymenium allows 'puffing' to occur. From the biological point of view *Taphrina deformans* also conforms to this type, with an extensive layer of closely-packed parallel asci displayed over the entire surface of the diseased part of the leaf of an affected peach or almond tree.

In the *Pyrenomycete* type the asci are contained in a small flask-shaped structure (whether perithecium or pseudothecium) which opens to the outside by a minute ostiole. Before each ascus can discharge the spores, its tip must reach the ostiole, the neck canal being so narrow that usually only one ascus can emerge at a time. Puffing is thus out of the question.

In the *Erysiphales* type the fruit-body is a cleistothecium somewhat like a minute perithecium but completely closed having no ostiole. In

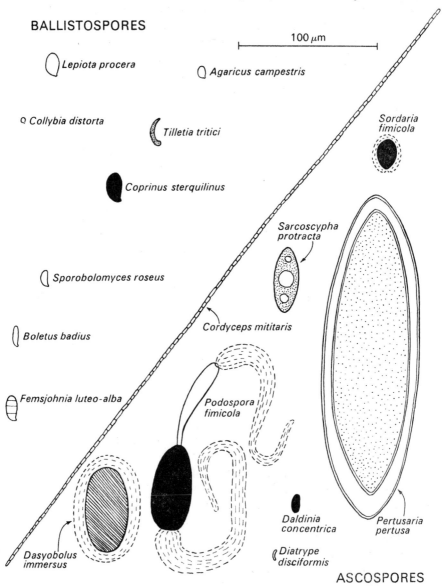

FIG. 2.2. Ballistospores (to left of long *Cordyceps* ascospore), and violently discharged ascospores. Illustrates range of form and size in the two types of spore.

this the swelling asci must first rupture the cleistothecium wall before they are in a position to discharge their spores.

In the *Myriangium* type, although the hymenium is exposed in a structure like a small apothecium, the almost spherical asci are embedded in a plectenchymatous tissue and are free to discharge only when this gradually undergoes gelatinization (Fig. 2.3). Species of *Myriangium* and

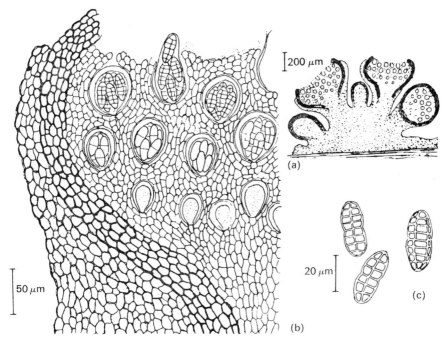

FIG. 2.3. *Myriangium duriaei*; (*a*) stroma (each loculate portion resembling an apothecium); (*b*) small portion more highly magnified; (*c*) ascospores. After von Arx (1968).

allied genera are essentially parasites of higher plants in the warmer regions and spore discharge from them has received little attention.

The first three types will now be further considered.

Discomycetes range from minute forms 0·2 mm in diameter (e.g. *Saccobolus* spp.) to the large morels (*Morchella* spp.). Many are more or less cup-shaped with the hymenium lining the inside of the cup (e.g. *Aleuria*, *Dasyscypha*, *Cookeina*, *Bulgaria*); in others there is a more or less horizontal or reflexed surface (e.g. *Rhizina*). Again, in some genera the hymenium is freely exposed as a stalked club as in *Geoglossum* and *Mitrula*. In the morels there is a stout stipe bearing an ovoid fertile head with the extensive hymenium lining irregular pits, the crests of the separating ridges alone being sterile.

The types of fruit-body architecture that have evolved seem to have depended on certain characteristics of the explosive ascus; first, that its spores are usually discharged to a distance of 0·5–2·0 cm; secondly, that the ascus is often positively phototrophic. A third point of importance is that the ascomycete hymenium is not easily harmed by rain, in contrast to that of Hymenomycetes which is completely disorganized when wetted.

Since ascospores are normally discharged to a distance of a centimetre or more, there is a reasonable chance of their effective dispersal when they are shot upwards. In immediate contact with the ground is a layer of air which is still or, if in motion, its flow is 'laminar'. Microscopic particles in this layer sink rapidly to the ground. Commonly it is very thin, perhaps only a millimetre or two deep, although sometimes it may become very much deeper, especially at night in settled weather. Thus discharged ascospores stand a good chance of being shot through the lower layer into the air above, which is likely to be turbulent, allowing spores to be borne aloft with a better chance of reasonable dispersal. In general, it may be said that Ascomycetes shoot their spores into the turbulent air while toadstools and bracket fungi drop their spores into this layer (Fig. 2.4).

Thus the roughly horizontal upward-facing type of hymenium is quite consistent with efficient dispersal and is, indeed, frequently developed. By contrast, the upward-facing hymenium hardly ever occurs in Hymenomycetes (toadstools and bracket polypores) where the spores are rarely shot from their basidia to a distance exceeding 0·02 cm and where their hymenia are so susceptible to injury by rain. On the other hand, the polypore or toadstool type of fruit-body in which vertical hymenial surfaces are closely opposed, could hardly have developed in Discomycetes without a very considerable decrease in the violence of ascus discharge.

Positive phototropism is probably a general feature of elongated asci (Buller 1934) although it has so far been reported only in Pezizales. A hymenium lining a deep narrow cup, as in *Sarcoscypha protracta* or the alveoli of *Morchella*, can function efficiently only because of this phototropism, for, as Buller has insisted, if the asci were simply arranged at right angles to the general surface of the hymenium, a high proportion of spores would be wasted by being flung against an opposite hymenial surface within range and directly in line of fire of the asci. The direction of discharge of the ascus jet is clearly determined by the direction in which the tip of the ascus points.

As an introduction to the consideration of spore liberation in Discomycetes, attention may first be focused on the genus *Ascobolus* and its close allies, nearly all of which are coprophilous.

Apart from its small size *A. stercorarius* is a fairly typical discomycete. It develops commonly if horse dung is kept for a week or two in a closed but ventilated container in the light without allowing the dung to dry or become excessively wet. The apothecium, usually about 1 mm in

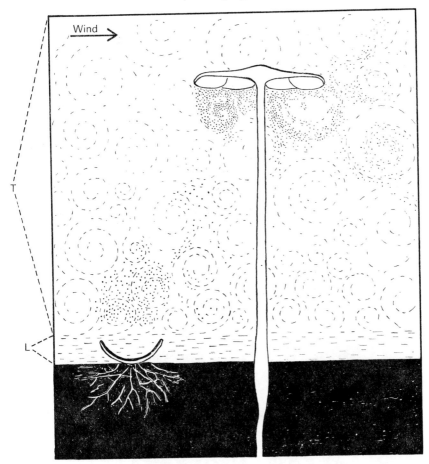

FIG. 2.4. A cup-fungus (*Peziza* sp.) and an agaric (*Oudemansiella radicata*). The ground is shown black. Above this is the laminar air (horizontal dashes) and above this the turbulent air (dashes in spirals). The *Peziza* has just discharged a puff of spores through the laminar air into the turbulent region. From the pileus of the agaric spores are steadily dropping into the turbulent air.

diameter, is clearly illustrated in Corner's beautiful drawing (Fig. 2.5). As in all Discomycetes, it is composed of two intergrowing systems of hyphae. One, consisting of ascogenous hyphae, arises from the fertilized female organ (ascogonium). Asci are formed from the ends of these hyphae, their mode of ultimate branching producing asci in a continuous succession and all at about the same level. The second hyphal system is of vegetative hyphae that grow up from below the ascogonium and form most of the apothecial tissue including the long, narrow paraphyses that are intermixed with the asci. The hymenium, consisting of asci and paraphyses, is devoid of air spaces, for mucilage occurs between its elements and

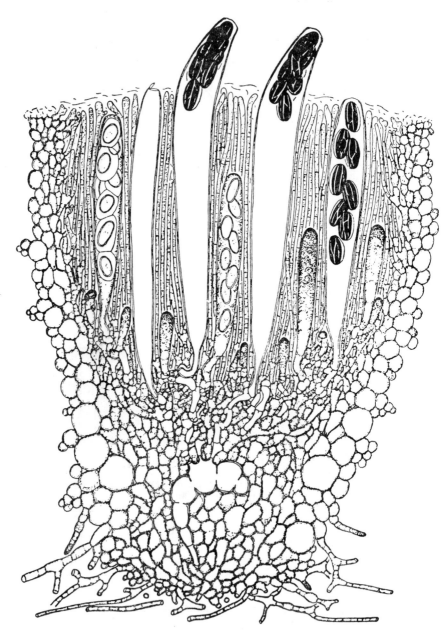

Fig. 2.5. *Ascobolus stercorarius.* Vertical section through apothecium. The projecting asci are pointing towards the incident light. × 300. After Corner (1929).

spreads out on the exposed surface as a thin layer pierced only by the protruding tips of the ripe asci which, being positively phototropic, point toward the light. The whole hymenium has a 'water-soaked' appearance.

In *Ascobolus stercorarius*, and in other species of Pezizales (operculate Discomycetes), when the ascus bursts a small apical lid hinges backwards, the stretched ascus-wall contracts longitudinally and transversely, and the spores, together with some ascus sap, are squirted into the air to a distance of several centimetres. As only a minute fraction of a second is involved the spores seem to escape simultaneously. However, since the orifice through which they leave the ascus is not so wide as the spores themselves, it seems that they must in fact escape in succession.

The most familiar of the Discomycetes are species with cup-like apothecia one to several centimetres in diameter. *Aleuria vesiculosa*, *Sarcoscypha coccinea*, *Galactinia badia*, and *Peziza aurantia* are familiar British examples. In Africa perhaps the commonest and most conspicuous cup-fungi are *Cookeina sulcipes* and *C. tricholoma*.

The organization of the cup-shaped apothecium in connection with spore liberation has been considered in detail by Buller (1934). In *Aleuria vesiculosa* (Fig. 2.6) it is sessile and lined by hymenium containing

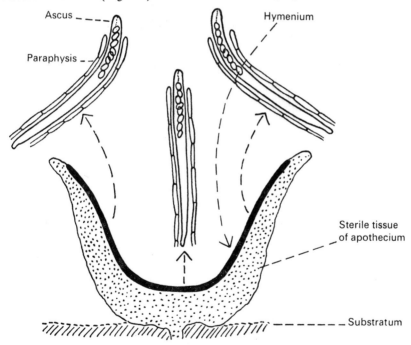

Ascus

Hymenium

Paraphysis

Sterile tissue of apothecium

Substratum

FIG. 2.6. *Aleuria vesiculosa*. Apothecium in vertical section. Hymenium shown black. An ascus and two paraphyses are shown taken from three indicated positions in the hymenium. Based on figures by Buller (1934). Apothecium ×4; asci ×150.

millions of elongated asci with filamentous paraphyses between. All elements of the hymenium are positively phototrophic and therefore point in the general direction of the mouth of the cup. Because of this response, asci and paraphyses from the bottom of the cup are straight, but those from its more vertical sides are strongly curved. In this species the eight ascospores form a single series, being attached to one another and to the apex of the ascus by a fine granular thread. When the ascus bursts a hinged lid is thrown back and the spores are shot into the air to a height of 1–2 cm, separating from one another in the process.

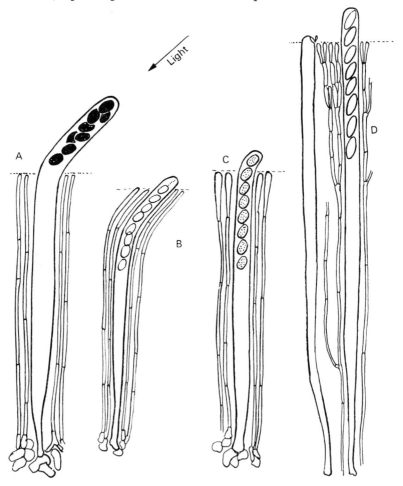

FIG. 2.7. Phototropism of asci. (A) *Ascobolus magnificus*, emergent part of ascus bends towards light. (B) *Aleuria vesiculosa*, both asci and paraphyses bend toward light. (C) *Scutellinia scutellata*, emergent tip of ascus turns to the light (D) *Sarcoscypha protracta*, no apparent ascus curvature, but lid displaced to more illuminated side of apex. All highly magnified. Based on figures by Buller (1934).

In *A. vesiculosa* all elements of the hymenium are phototrophic, but this seems to be exceptional and usually response to light is limited to the projecting parts of the ripe asci as in *Ascobolus* or in *Scutellinia scutellata* where the ripe asci project only slightly (Fig. 2.7).

In *Sarcosypha protracta* (Fig. 2.8) there is no *evident* response even in the projecting parts of the mature asci. It is, however, argued that curvature does actually occur but is limited to the extreme tip. The result is that the lid is displaced towards the more strongly illuminated side of the apex, but this is sufficient to direct the issuing ascus-jet towards the incident light (Figs. 2.7 and 2.8).

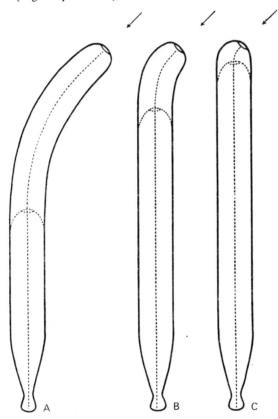

FIG. 2.8. Phototropism of asci. All asci initially straight when position of apex was as shown by the broken line. In *Aleuria* (A) curvature starts when ascus is only half its final length. In *Scutellinia* (B) it occurs later and in *Sarcoscypha* (C) still later. Axis of ascus shown by dotted line. Arrows indicate direction of light. After Buller (1934).

The situation that Buller has described in *S. protracta* also obtains in species of the common tropical genus *Cookeina* (Fig. 2.9). The asci are all quite straight, but the lid is produced on the brighter side of the apex facing the mouth of the apothecial cup. A feature of special interest is that only a single crop of asci is produced, a striking contrast to the continuous formation of asci spread over many days or weeks which is the normal condition in the hymenium of a cup-fungus. Since in *Cookeina*

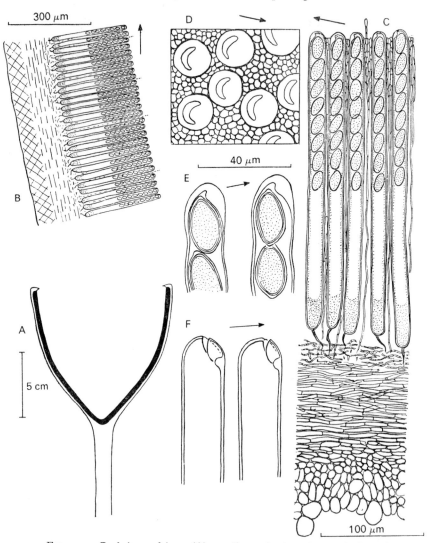

Fig. 2.9. *Cookeina sulcipes*. (A) small apothecium in vertical section; hymenium as very thick black line. (B) part of nearly vertical region of A more highly magnified. Asci shown in actual form, but other tissues indicated diagrammatically. (C) part of B shown in detail. (D) surface view of small portion of hymenium drawn at a focal plane about 5 μm behind extreme apex of asci. (E) the apieces of two mature asci. (F) apical parts of two dehisced asci. The arrows in all cases point towards the mouth of the apothecium.

all the asci are exactly at the same stage of development, the number of asci produced per square centimetre of the hymenium can be accurately determined from tangential surface sections of the hymenium. This works out at 215 000, involving a spore production of 1 720 000. The beautifully regular, stalked apothecia, each like a little wine-glass, are 1–3 cm in diameter and it has been calculated that the spore production from each varies, depending on size, from 3 to 24 million spores (Ingold 1966).

A feature of most apothecia is the phenomenon of 'puffing'. This has been observed not only in a considerable range of operculate Discomycetes but also in a number of inoperculate genera (e.g. *Sclerotinia*, *Mollisia*, and *Rhytisma*). Puffing, noted as early as 1729 by Micheli in *Novum plantarum genera* (Fig. 2.10B), was considered in some detail by

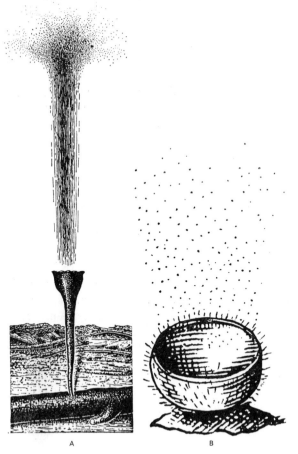

FIG. 2.10. (A) Buller's (1934) figure of puffing in *Sarcoscypha protracta*. (B) Micheli's (1729) figure of the same process in a small cup-fungus.

Buller (1934). It involves the simultaneous bursting of a very large number of asci with the liberation of a visible smoke-like cloud of spores. If an apothecium is held close to the ear at the moment of puffing a hissing may be heard. In most cup-fungi the sound lasts only for a second or two, but in *Rhizina inflata* it may go for several minutes, gradually dying away.

Buller (1934) gave special attention to puffing in *Sarcoscypha protracta*. In this, because of the phototropism of the asci (limited, however, to their extreme tips), the spores are shot outwards on puffing as a parallel beam (Fig. 2.10A), rather as the parabolic mirror of a searchlight throws a parallel shaft of light into the night sky. One such spore-beam was reported to have travelled 12 cm and another 17 cm, while the distance of throw for asci discharging singly was only a few centimetres. He suggested that, on puffing, the air above an apothecium is set in general motion, with the result that the spores travel considerably further than would be possible if asci burst singly. Buller also demonstrated that a detectable blast of air is produced when the apothecium of *S. protracta* puffs.

Puffing is also a striking feature of *Ascobolus*. If an undisturbed, fruiting, Petri-dish culture of *A. crenulatus* is examined, the individual apothecia appear dark purple (almost black) due to the projecting ripe asci with their pigmented spores. If the lid of the dish is suddenly removed the apothecia almost immediately puff, and instantaneously they become pale greenish-yellow.

Individual apothecia of *A. crenulatus* have been used in a laboratory study of spore discharge with special reference to puffing (Ingold and Oso 1969). However, before discussing this matter further, something should be said about how the course of spore liberation may be followed. In the work on spore discharge in *Ascobolus* and *Sordaria* considered in this chapter, what has been called a 'spore clock' has mostly been used (Ingold and Marshal 1963). Its essential structure is illustrated in Fig. 2.11. A very shallow box of black opaque Perspex, extended below into a cubicle to accommodate the discharging culture, has rotating horizontally in it a disk of transparent Perspex. This, attached to the vertical axis of a clockwork mechanism, completes a single rotation in a definite time, often adjusted to just over 24 hours. Spores are discharged onto the under surface of the disk, which has a grid etched on it to facilitate counting of the spores under the microscope. In the lid of the box above the cubicle is a small window through which the fungus can be illuminated. If, however, this is blacked-out, the interior is in darkness. The whole lid is detachable, but when in position it can be firmly secured by screws. Entry and exit tubes, near the top of the cubicle, allow an air stream to be passed over the culture if desired. At the end of an experiment the disk can be removed and the hourly rates of discharge estimated

by suitable spore-counts under the microscope. In work on spore discharge it is often most convenient to use cultures grown in glass specimen
tubes ($2\cdot5 \times 2\cdot5$ cm) brimfull with the appropriate medium.

It has been found that at 20°C a single apothecium may continue to
liberate its spores over a period of 3 weeks. In some species of *Ascobolus*,
discharge seems to be limited to a daily puff, but in *A. crenulatus*, although
a number of treatments cause puffing, there is a steady 'background'
discharge that goes on fairly continuously except for a few hours immediately following a puff. In continuous darkness, in nearly saturated
air, there is a steady discharge, but this is maintained at a relatively
low level. However, after a day or two in darkness an apothecium can
regularly be caused to puff by a number of treatments, especially illumination with blue light of sufficient intensity, sudden subjection to air
of low relative humidity, transfer to a low temperature (e.g. 3–5°C),
or touching with a needle.

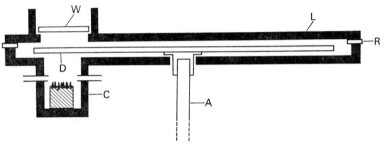

FIG. 2.11. Sectional view of spore-clock; one third actual size. Black
Perspex indicated by very thick black lines. (L) lid of box; (R) rubber
washer; (W) light filter; (C) cubicle containing culture and fitted with
inlet and outlet tubes; (D) Perspex disk; (A) vertical axis connecting to
clock (not shown).

In *Ascobolus* a change of conditions may lead to a dramatic increase in
the density of the spore deposit on the moving disk of the spore-clock
above the discharging apothecium, but it cannot be certain that this is
produced as a result of a puff. However, when such an increase is followed
by a fall to essentially zero for several hours, it is highly probable that a
puff has occurred.

With an intensity in excess of 100 lx, light at the blue end of the spectrum (400–460 nm) regularly causes puffing to occur in an apothecium
that has previously been kept for several hours in darkness (Fig. 2.12),
but light of longer wavelengths (500–750 nm) has no such effect (Fig.
2.13). On transfer from darkness to white light no puffing results, but
over a period of several hours the rate of spore discharge steadily climbs.

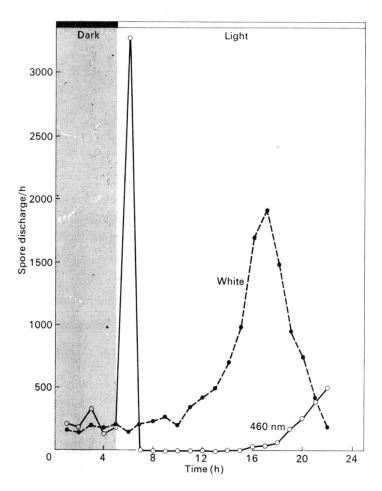

FIG. 2.12. Hourly rate of discharge plotted against time for two apothecia of *Ascobolus crenulatus*. Prior to start of experiment each apothecium was in darkness for 2 days. Five hours after start one (unbroken line, white dots) was illuminated with light of 460 nm wavelength; the other (broken line, black dots) with white light. Intensity of both 500 lx; temp 20°C; continuous flow of saturated air over apothecia.

Moreover, when an apothecium, conditioned in darkness, is illuminated with blue light sufficiently intense to be effective, and at the same time with light of similar intensity but above 600 nm in wavelength, no puffing occurs. It seems that this light towards the red end of the spectrum can counteract the stimulation of the blue light (Fig. 2.13B).

Response to alteration in temperature is complex. A change from 19°C to 30°C, or the reverse, seems to have little effect on the rate of

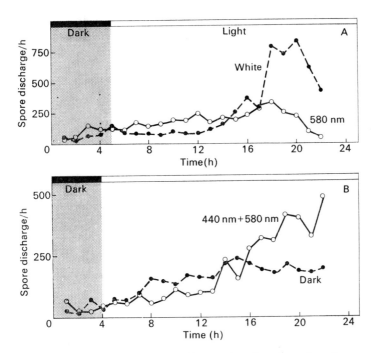

FIG. 2.13. (A) Hourly rate of discharge plotted against time for two apothecia of *Ascobolus crenulatus*. Prior to start of experiment each apothecium was in darkness for 2 days. Five hours after start one (unbroken line, white dots) illuminated with light of 580 nm wavelength 500 lx; the other (broken line black dots) with white light. (B) similar experiment one apothecium (broken line, black dots) being retained in darkness; the other (unbroken line, white dots) receiving, after 4 hours in darkness, light of 440 nm wavelength plus light of 580 nm wavelength both at an intensity of 500 lx. For both experiments temp 20°C and continuous flow of saturated air over apothecia.

discharge in the dark. However, a change from a relatively high temperature to a very low one (3–4°C) tends to result in a puff (Fig. 2.14A).

In experiments in which the apothecium could be rapidly and vigorously vibrated, it was found that transition from rest to a state of vibration caused no puffing to occur. If an apothecium with ripe asci was actually touched, however, puffing invariably occurred (Oso 1969a).

Reference has already been made to the increased distance of discharge as a result of puffing. This has been measured objectively in *A. crenulatus*, taking advantage of the fact that steady discharge continues in darkness but, after a prolonged period in the dark, puffing can be induced by blue light (Ingold and Oso 1968). An apothecium was allowed to discharge horizontally slightly above the level of a horizontal glass collecting-slide in an elongated dark box. The spores fell onto this slide

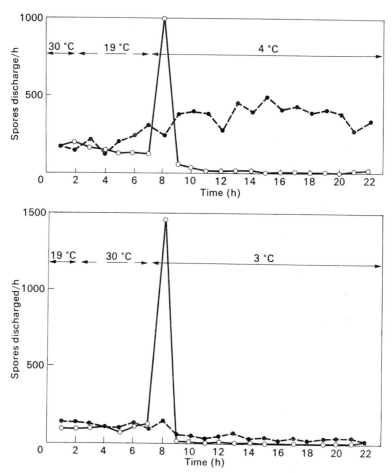

FIG. 2.14. Effect of changes of temperature on spore discharge from *Ascobolus crenulatus* in darkness. *Above*, hourly rate of discharge plotted against time. Prior to start apothecium at 30°C in dark for 2 days, following which record taken over 2-h period at 30°C before transferring to 19°C for 5 h and finally to 4°C. Control (dashed line) at 30°C throughout. *Below*, prior to start of experiment apothecium at 19°C for 2 days in dark, then record of discharge over 2-h period at 19°C before transferring to 30°C for 5 h and finally to 3°C. Control (dashed line) is at 3°C throughout.

and their distance from the hymenium could later be determined. A window at the end of the box could either be blacked out or, when desired, the apothecium could be illuminated through it by a horizontal beam of blue light. Spore deposits obtained during many hours of darkness alone, or followed by brief treatment with blue light, were compared. It was found that with the addition of the puff triggered by the blue light, the lateral spread of the spores in the deposit was greatly increased

(Fig. 2.15). In *Ascobolus crenulatus* most of the projectiles are single spores, but there are also groups of from two to eight. Since with a given initial velocity the distance of discharge increases with size of projectile, the comparison of distance of discharge has been confined to single spores (Fig. 2.15).

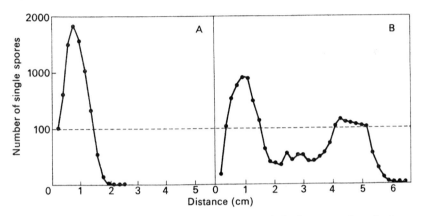

FIG. 2.15. *A. crenulatus*. Record of numbers of single spores deposited from apothecia discharging horizontally plotted against distance from apothecium. Scale above 100 (broken line) reduced to one-tenth. (A) deposit accumulated in darkness for 48 h. (B) deposit accumulated in darkness for 48 h followed by 1 h of blue light.

In the course of studies on spore dispersal in Discomycetes, Falck (1948) considered that two types could be recognized. In the first, the radiosenstive type, puffing does not occur, but spore discharge is greatly stimulated by incident radiation. Falck considered that *Morchella* and *Gyromitra* are of this type. In the second, the tactiosensitive type, puffing occurs in response to mild shocks, as when an apothecium is touched or blown upon. Many genera belong to this type, including *Humaria*, *Peziza*, and *Geoglossum*. But it is doubtful if the distinction is valid. Thus *Ascobolus crenulatus* is clearly both radiosensitive and tactiosensitive, while *Ascophanus carneus* seems to belong to neither type. *Bulgaria inquinans*, nevertheless, does appear to be radiosensitive in Falck's sense, discharging its spores much more actively in light than in darkness, but not apparently given to puffing.

Among Discomycetes the morels (*Morchella* spp.) have the most elaborate kind of apothecium. There is a stout sterile stalk and the spore-producing layer lines a number of alveoli or shallow pits. The hymenium of asci and paraphyses is continuous except for the sterile crests of the major ridges separating neighbouring pits. The asci are phototropic and thus the escape of spores from the alveoli is assured (Fig. 2.16).

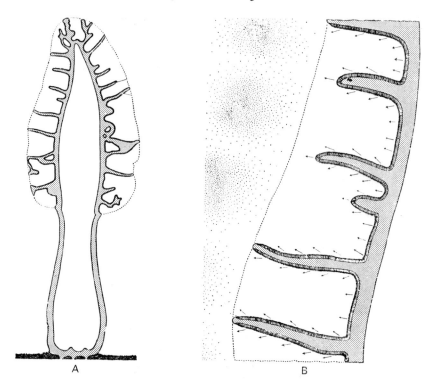

A   B

FIG. 2.16. *Morchella conica*. Vertical section of fruit-body A×1, B×4.
Dotted line shows general outline of fertile cap. B shows small part of A.
Note hymenium covering cap except for edges of major ridges. Discharged
spores indicated reaching that position because of phototropism of asci.
After Buller 1934.

Morchella has been regarded as the discomycete equivalent of the
hymenomycete *Boletus*. In both the hymenium lines pores, the striking
difference in size of these being associated with the difference in range of
ascus and basidium. In *Morchella* it is the phototropism of the asci that
determines the escape of the spores from the pits, which are mainly
horizontal; in *Boletus* it is geotropic response, leading to the verticality
of the downward-facing hymenium tubes, which is so important in
allowing the spores discharged into the middle of the tubes to fall freely
until they reach the air below the cap.

In the Discomycetes considered so far the spores escape from the
ascus in such a minute fraction of a second that discharge appears to be
simultaneous. However, in a few Discomycetes successive escape of the
spores is clearly to be seen, especially in Geoglossaceae.

The apothecium of *Trichoglossum hirsutum* or of *Geoglossum ophioglos-
soides*, species often seen on lawns in autumn, is black, club-shaped, and
5–10 cm high, with the hymenium covering the upper half of the club.

The ascus, which contains a sheaf of eight long, septate ascospores (Fig. 2.1), dehisces by a minute apical pore. As soon as this is formed one of the spores is forced into it, temporarily stoppering the ascus. At first this spore is slowly pushed out of the ascus by the hydrostatic pressures within until about half is projecting. Then it rapidly gathers speed and finally moves so rapidly that its flight cannot be followed by the eye until, slowed down by friction with the air, it flashes into view about half a centimetre from the hymenium. Immediately one spore has been discharged another takes its place, blocking the pore before the ascus can undergo any visible shrinkage. This in turn is shot away a few seconds after the first, and so on until the whole complement of eight has been liberated. As soon as the last spore has been ejected the ascus suddenly contracts, since there is no longer a spore to act as a temporary stopper.

Certain hypogeal Ascomycetes, in which spore discharge no longer occurs and dispersal is by rodents, seem to have had apothecial ancestors. The clearest cases are to be seen in *Genea*. In species of that genus the ascocarps are several centimetres below the ground; the asci retain their elongated form and are packed in a palisade-like hymenium interspersed with paraphyses (Fig. 2.17). These, however, unite above the level of the asci into a firm pseudoparenchymatous tissue. Further, the fruit-body is almost closed. In species of truffle classified in the genus *Tuber* the relationship to apothecial forms is by no means so clear.

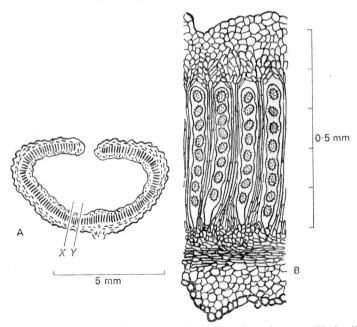

Fig. 2.17. *Genea hispidula*. (A) Vertical section of apothecium. (B) details of part of A between the lines X and Y.

Moreover the asci are almost spherical and not organized in definite hymenia.

Such a fungus as *Geopora sepulta* seems to suggest a step toward the hypogeal habit. The apothecium is at first nearly buried, with a small opening just above ground level. Only when it opens out is the hymenium fully exposed in a manner that allows free spore discharge.

Further, a very curious state of affairs has been reported by Burdsall (1965) in the subterranean sporophore of another species of the same genus, *G. cooperi*. It was found that on breaking open a fresh specimen a considerable area of convoluted hymenium was exposed from which puffing occurred from asci with well-defined opercula. Perhaps this fungus is a very recent recruit to the underground mycoflora (Fig. 2.18).

In the majority of Ascomycetes the asci are contained within a tiny flask-like structure. This will be referred to as a perithecium, although technically a distinction should be made between a perithecium and a pseudothecium. However, both are essentially similar biologically and even on grounds of morphology they can be distinguished from one another only by careful study of the course of development.

The perithecium opens by a narrow canal passing through the neck which may be short (e.g. *Sordaria*) or quite long (e.g. *Ceratostomella*). The canal is usually lined by filamentous periphyses but not in *Ceratostomella* spp. For successful liberation the tip of the ascus must project beyond

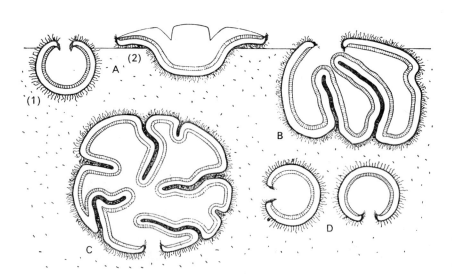

Fig. 2.18. Schematic drawings of *Geopora* spp. showing orientation of the hymenium and opening of the ascocarp. Horizontal line shows upper boundary of the soil (dotted). (A) *Geopora sepulta* before (1) and after (2) opening completely. (B) *G. longii*. (D) *G. clausa*, showing random orientation of opening. (C) *G. cooperi* with basal opening. After Burdsall (1968).

the mouth (ostiole) of the perithecium before it explodes. Owing to the narrowness of the neck-canal only one ascus can, as a rule, reach the ostiole at a time, so that simultaneous bursting of asci, so characteristic of Discomycetes, cannot occur.

Among the flask-fungi, species of Sordariaceae have been most widely studied. For actually watching the process of discharge some common species of *Podospora* are especially suitable. In *P. minuta* and *P. curvula* the wall of the perithecium is semi-transparent, so by focusing an optical section the behaviour of the living asci can be seen in a specimen mounted in water (Fig. 2.19). Most of the asci occur crowded in the lower half of the perithecium attached to a basal cushion of tissue, but the upper half is occupied by the greatly swollen upper parts of those asci that are fully mature. Between the perithecial wall and the mass of asci are thin-walled much inflated cells of the paraphyses. The top of the perithecial cavity tapers and finally merges into a short, narrow canal leading to the outside. One of the ripe asci in the upper region of the perithecium is normally slightly in advance of the others, and this may be seen to elongate, in part probably due to actual growth and in part to the pressure of the surrounding turgid cells. Finally, the tip of the ascus reaches the ostiole and protrudes slightly beyond it. It then bursts, discharging a projectile consisting of the ascus cap, eight spores roped together by strands of mucilage, and some ascus sap. On discharge the wall of the ascus, which is greatly stretched but is still attached, retracts into the perithecium where it quickly becomes disorganized. As soon as one ascus has discharged, another begins to elongate up the neck canal, and so the process goes on in orderly succession. It should be noted that in an active perithecium no gas phase is present, any space between the hymenial elements being occupied by mucilage. In sordariaceous fungi packing is so tight that there is virtually no free space, but in some Pyrenomycetes a considerable amount of mucilage occurs, enveloping the asci and paraphyses within the perithecium.

A feature of *Podospora*, *Sordaria*, and a number of non-stromatal flask-fungi is that the neck is positively phototropic.

Sordaria fimicola has been used extensively in the study of spore discharge. The structure of the perithecium is illustrated in Fig. 2.20. The course of spore discharge closely resembles that of *Podospora* but the ascus remains cylindrical throughout its development (Fig. 2.21). Moreover, it dehisces by an apical pore only 4 μm across. Each ascospore is often about 22×13 μm, although the size depends on the particular isolate and on the conditions of culture. Around each ascospore is a layer of mucilage about 3 μm thick.

Because the pore has a diameter much less than that of the spore, it must be considerably stretched during discharge. Contraction will tend to occur after the escape of each spore and before the pore begins

F<small>IG</small>. 2.19. *Podospora curvula*. (A) perithecium with the asci showing through the semi-transparent wall; (B) ostiole region of the same perithecium 10 min later—the tip of the leading ascus in A has now reached the outside and is about to discharge its spores; (C) a single ascus that has just commenced to swell; (D) a discharged spore-mass showing the ascus cap, partly turned inside out, attached to the apical spore. A and B×83, C and D×270.

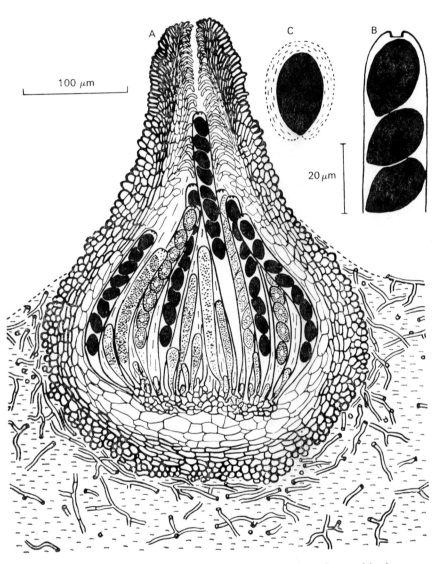

FIG. 2.20. *Sordaria fimicola.* (A) Vertical section through a perithecium growing on agar (dashed lines); (B) Vertical section of ascus tip (protoplasmic lining not shown); (C) single liberated spore covered with mucilage.

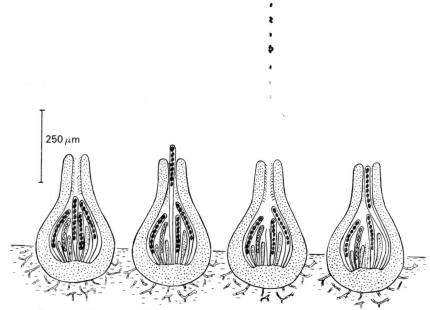

250 μm

FIG. 2.21. Diagram of discharge in *Sordaria fimicola*. On the left the tip of an ascus is just entering the neck canal; (a few minutes later the tip of the ascus protrudes beyond the ostiole; the next instant the ascus has burst and the empty envelope has retracted into the perithecium; some minutes later the old empty ascus is breaking down and another ascus is elongating up the neck canal.

to be stretched again by the next one in the row. On account of their mucilaginous sheaths, successive spores tend to stick together, but the contracting pore may break this connection. Thus in a spore deposit formed from perithecia discharging horizontally, spores occur singly and also in groups of two to eight.

This matter has been studied (Ingold and Hadland 1959) using cultures in specimen tubes ($2\cdot5 \times 2\cdot5$ cm) grown with overhead lighting so that the phototropic necks are parallel with the longitudinal axis of the culture tube. Such a culture was placed on its side in a long black box (Fig. 2.22) and illuminated by a horizontal beam of light. Spores discharged horizontally fell onto a graduated glass slide. After a suitable period the slide was examined microscopically and the numbers and positions of all the examples of each size of spore group (1 up to 8) were determined (Fig. 2.23). It is clear that the largest number of projectiles were 1-spored, thereafter the number fell to the 7-spored type, but rose again for the 8-spored. It can be shown that this is often to be expected if the chance of a break in the chain of eight escaping spores is equal at each of the seven links in the chain.

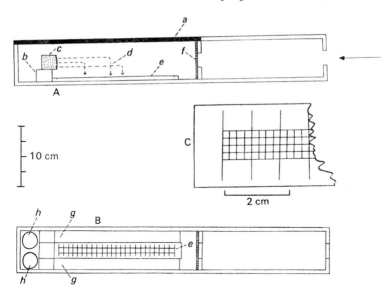

FIG. 2.22. Wooden box used in studying horizontal discharge in *Sordaria*. (A) vertical section of box: (*a*) removable lid; (*b*) culture holder; (*c*) specimen-tube culture; (*d*) trajectories of discharged spores; (*e*) graduated slide; (*f*) glass window. The arrow indicates the incident light. (B) plan of box; (*e*) graduated slide; (*g*) wooden guides holding slide in position; (*h*) specimen tubes containing water. (C) small part of graduated slide at larger scale to show details.

The tendency for spores to separate (θ) can be calculated from the expression

$$\theta = \frac{N - \frac{1}{8}n}{\frac{7}{8}n},$$

where N is the total number of groups of spores and n the total number of spores. It follows that the number of links between spores is $\frac{7}{8}n$. If the spores always separate from one another (i.e. $N = n$) the value of θ is 1·0; if they always stick together θ is zero.

In the experiment illustrated in Fig. 2.23, θ is 0·168. It has been remarked that 8-spored groups considerably exceed the number of 7-spored groups. It can be shown mathematically that the number decreases from 1-spored to 7-spored groups and then rises or falls so far as the 8-spored group is concerned according as θ is less than or exceeds 0·3. Figure 2.24 shows the actual number of groups of each size, and also the theoretical curve based on the view that a break is equally likely between any two escaping spores. Agreement between observation and theory is clearly good.

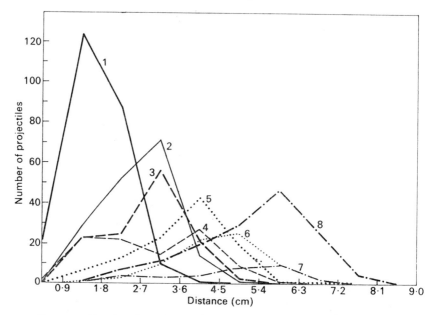

Fig. 2.23. Graphs of number of projectiles in each major square (0·9 × 0·9 cm) of graduated slide (see Fig. 2.22) plotted against distance from culture for each of the eight sizes of projectile. The figure associated with each graph indicates spores per projectile. This is Expt. 1 (see Fig. 2.24).

No doubt the tendency for spores to stick together is affected by the viscosity of the mucilage around the spore. Since increase in temperature lowers viscosity, it might be reasonable to suppose that the θ value would increase with temperature. This has proved to be so. In one experiment θ at 21–24°C was 0·252 and at 7–10°C, 0·183; in another θ was 0·288 at the higher and 0·227 at the lower temperature (Ingold and Hadland 1959).

A consideration of the distribution by size of spore-groups in a deposit formed by the settling of ascospores of *Sordaria* discharged horizontally leads to information about the association of spores during discharge or, stated more generally, about the behaviour of the ascus jet. Another way of obtaining information is to catch jets on the underside of a Perspex disk rotating very rapidly in a horizontal plane just above discharging perithecia (Fig. 2.25).

If the speed of rotation is sufficient the jet discharged from each ascus is spread out horizontally on the disk and can subsequently be observed under the microscope. It can be seen (Fig. 2.26) that in such impaled jets there are not only spores in groups of various sizes, but also ascus sap either associated directly with spores or as isolated droplets. It has generally been supposed that in some Ascomycetes the spores of an ascus

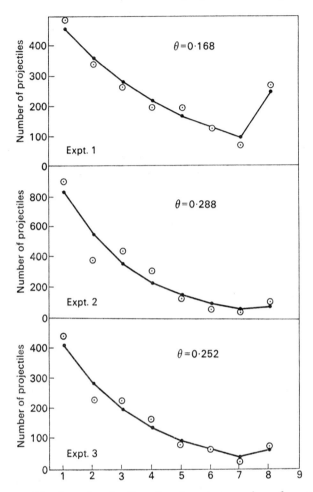

Fɪɢ. 2.24. Number of projectiles plotted against number of spores per projectile (Expt. 1 is that referred to in Fig. 2.23). The solid line connects the theoretical values calculated from the given value of θ and the total number of spores involved.

separate from one another on discharge while in others they always stick together. This may be roughly true, but all types seem to exist from species where the spores invariably stick together, as in *Saccobolus* spp. and *Dasyobolus immersus*, to those in which they always separate. It has been suggested that the former condition is especially a feature of coprophilous species, but there are certainly non-coprophilous fungi, such as *Eutypa armeniacae* (a parasite of apricot shoots in Australia) and the discolichen *Lecanora conizioides*, in which the ascus complement of eight spores normally forms a single projectile. In the lignicolous

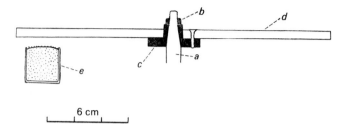

6 cm

FIG. 2.25. Sectional view of apparatus used to study the form of ascus jets: (*a*) spindle of electric motor; (*b*) nut; (*c*) brass socket fitting spindle; (*d*) Perspex disk; (*e*) fruiting culture of *Sordaria*.

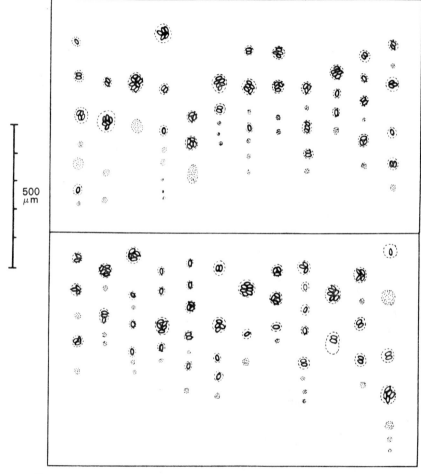

500 μm

FIG. 2.26. *Sordaria fimicola. Camera lucida* drawings of twenty-four ascus jets caught on the under surface of a rotating disk. In all the front end of the jet is uppermost. Rate of movement of the part of the disk above the discharging perithecia was 1672 cm/s.

Daldinia concentrica, as in *Sordaria*, all types of projectile from 1-spored to 8-spored are formed with, however, a strong tendency for the spores to separate expressed by a θ value of 0·40.

Although in the initial episode of dispersal in *D. immersus* and in *E. armeniacae* the spores are in octads, this does not necessarily mean that they will remain together to the end of the dispersal process. Thus in the former, which is coprophilous, the spores probably separate during passage through the alimentary canal of a herbivore; and in the latter the spores, initially deposited on the branch of an apricot bush may, perhaps, be splashed and separated by a falling raindrop in a further phase of transport (Carter 1957).

In *S. fimicola* spores are discharged horizontally up to 10 cm and it can be shown that, for the size of projectile involved, the vertical throw would be only slightly less. The larger projectiles are shot on the average to a greater distance than the smaller ones (Ingold and Hadland 1959). With a given muzzle velocity of the ascus gun the distance D to which a spherical projectile is shot can be roughly expressed by $D = Kr^2$ (see p. 6). Thus assuming that all *Sordaria* projectiles, irrespective of the number of spores, have the same take-off velocity and that they are spherical, we should expect the larger ones to carry further in accordance with the equation. However, the average distances actually observed for projectiles of more than one spore are less than expected (Fig. 2.27), but if a correction for Reynold's number is made the agreement is improved. Walkey and Harvey (1966), using *S. macrospora*, possibly to be regarded merely as a larger-spored variety of *S. fimicola*, have confirmed this relationship, but they could find no evidence of it in other fungi that they studied. Indeed, in *Xylosphaera longipes* and *Ascobolus stercorarius* the average distance of discharge decreased slightly with increase in the number, from one to eight, of spores in the projectile. Nevertheless in *Daldina* the general relationship holds (Ingold 1956*b*). If a median vertical slice, about half a centimetre thick, is cut from an active stroma and laid horizontally on a glass plate, a dense spore deposit accumulates overnight. Instead of this being a somewhat blurred line, it takes the form of a blackish band over 1 cm wide, starting about 0·5 cm from the surface of the stroma (Plate II). Microscopic examination shows that the spores nearest the stroma are mostly singletons while those farthest away are octads. Again the band is fairly clearly zoned with a number of darker and lighter regions parallel to the stromatal surface. If spore groups of increasing size (1–8) were shot to characteristic distances, eight dark zones might be expected. However, although this is clearly not what happens, the actual banding observed may reasonably be attributed to a tendency in this direction.

It may be appropriate here to refer to the general question of the importance of projectile size in relation to distance of discharge. In a study of

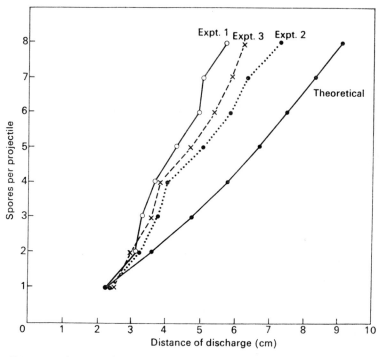

FIG. 2.27. Average distance of discharge for each of eight sizes of projectile plotted against number of spores per projectile. Results from three separate experiments. Theoretical curve base on $D = Kr^2$ with the value for K selected to give the observed distance for single spores in Expt. 1 (see Fig. 2.23).

seven Ascomycetes ranging from species with rather small spores to those with very large spores, it was claimed (Ingold 1961) that there tended to be reasonable agreement with the equation $D = Kr^2$; that is to say that size of projectile rather than differences in vigour of discharge seemed to be the master factor influencing distance of discharge. However, an investigation by Walker and Harvey (1966), covering a much larger range and number of species, failed to confirm this relationship. Nevertheless, it is clear that Ascomycetes that shoot their spores to relatively great distances have large spores, or have spores bound together in a single mass, or have both these features. For example, the 30-cm discharge distance of *Dasyobolus immersus* is associated with a projectile of eight big spores (each 60×35 μm) glued together in a single mass (Buller 1909) and the same is true of *Podospora fimicola* with spores measuring 54×37 μm (without considering their appendages) in which the 8-spored projectile may be thrown to a distance of 50 cm (Walker and Harvey 1966).

Although so far consideration has been given to the *Sordaria* type, which is probably the most general one among perithecial Ascomycetes, species with double-walled asci should receive some consideration. It is now considered that this 'bitunicate' ascus has great taxonomic value and that it characterizes the large order of Pseudosphaeriales. In these the

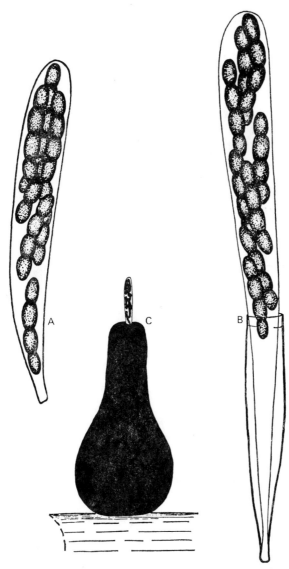

Fig. 2.28. *Sporormia intermedia*. (A) ascus before rupture of outer wall, ×700. (B) ascus after rupture of the outer wall and elongation of the inner wall, ×700. (C) pseudothecium on piece of straw in horse dung, with tip of ascus protruding, ×124.

perithecium (pseudothecium), although biologically like that of other 'Pyrenomycetes', has a different development.

An example is *Sporormia intermedia* (Fig. 2.28), a fungus often found with species of *Sordaria* and *Podospora* in the horse-dung flora. It also fruits readily in pure culture. The asci occur attached to a basal cushion within the perithecium, but each has a jack-in-the-box construction. The ascus wall consists of an outer rather rigid membrane and a very thin, inner, extensible one. Just before discharge the outer membrane is ruptured apically and, the wall pressure being now reduced, the ascus is free to absorb water, with the result that it elongates very rapidly up the neck canal of the perithecium and soon projects through the ostiole; it then bursts, squirting its spores into the air.

In the bitunicate ascus a constant feature is the enormous swelling of the thin inner wall that occurs immediately after discharge, due, no doubt, to the absorption of water. This was clearly illustrated by Pringsheim over a hundred years ago (Fig. 2.29).

Discharge in *Leptosphaeria acuta* (Fig. 2.30), common on dead stalks of nettle, is like that in *Sporormia*, but the spores, as in *Geoglossum*, are shot in succession through an apical pore.

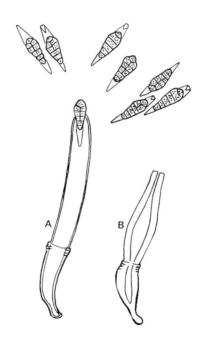

FIG. 2.29. *Pleospora scirpicola*. (A) Ascus in the act of discharging its eighth spore; the seven other spores are seen outside; (B) empty ascus immediately after discharge of the eighth spore. After Pringsheim (1858).

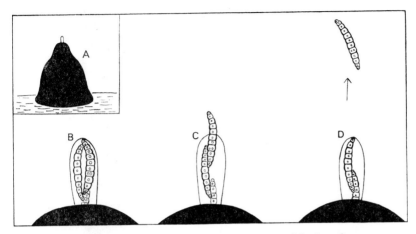

FIG. 2.30. *Leptosphaeria acuta*. (A) perithecium with tip of an ascus protruding; (B–D), apex of perithecium showing successive spore discharge from protruding ascus. (A) × 8, (B–D) × 200. Slightly modified after Hodgetts (1917).

Successive discharge of elongated spores in flask-fungi reaches its extreme expression in *Cordyceps*, in which, however, the ascus is unitunicate. *C. militaris* is a common parasite of the larvae and pupae of Lepidoptera. From a dead mummified pupa in the soil there grow above the ground orange, club-shaped stromata, the upper parts of which bear numerous perithecia projecting as conical pimples. In this fungus, as in Clavicipitales generally, the ascus is very long, narrow and cylindrical. Its tip is thickened and at dehiscence is pierced by a narrow tubular pore. The spores are long and narrow (400–500 × 2 μm) forming a sheaf of eight within the ascus. By laying a stroma on a slide and examining it under the microscope discharge can readily be observed. The tip of an ascus suddenly appears at an ostiole and rapidly elongates until it is protruding to the extent of 40–60 μm. Then suddenly an ascospore flashes into sight about 200 μm from the ostiole and quickly sinks from view. A second or two later, another appears, and so on. As soon as all eight spores are discharged, the ascus tip is no longer visible. After an interval, varying from a half to ten minutes, another ascus protrudes and discharges its spores (Fig. 2.31).

A curious feature of discharge in *Cordyceps* is that the spore leaves the ascus at such speed that nothing is to be seen of it until it suddenly appears about a quarter of a millimetre from the ostiole; at no stage is it seen protruding from the ascus and stoppering it as in *Geoglossum* or in *Leptosphaeria acuta*. It is difficult to understand how the long thread-like spore can reach sufficient speed within the ascus so as to leave it without being visible to an observer.

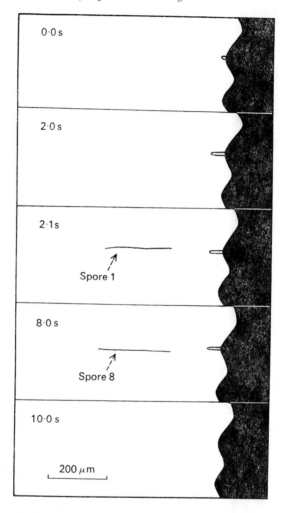

Fɪɢ. 2.31. *Cordyceps militaris.* Two projecting perithecia as seen when stroma is laid on its side. At 0·0 s the tip of an ascus is beginning to project from an ostiole. At 2·0 s the projecting tip has reached its maximum length. At 2·1 s the first spore flashes into view and then rapidly falls out of sight At 8·0 s the last spore has been discharged. At 10·0 s the empty ascus has retracted into the perithecium.

In *Epichloe typhina*, another member of Clavicipitales, discharge is much as in *Cordyceps*, but the interval between the liberation of successive spores is very small and usually the escape of all eight occupies less than a second. Sometimes a spore follows its predecessor so rapidly that its apex sticks to the base of the predecessor and a long filament consisting of several spores end to end may then be discharged in one piece.

It is interesting to consider the symmetry of elongated ascospores that are violently discharged (Ingold 1954). In the great majority the upper half, that nearest the apex of the ascus, is the mirror image of the lower. This type may be termed 'bipolar symmetrical'. However, in many species, distributed widely in both Discomycetes and Pyrenomycetes, an imaginary cut at the mid-point between the two ends of the spore would divide it into a relatively large and blunter upper part and a relatively small lower part with a sharper end. This is the 'bipolar asymmetrical' type (Fig. 2.32). The interesting fact is that the opposite condition, a

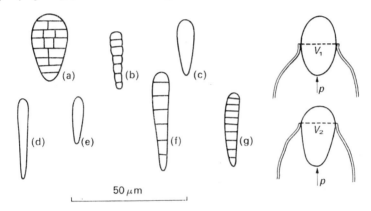

Fig. 2.32. *Left:* bipolar asymmetrical ascospores. Lichens: (*a*) *Arthothelium spectabile*, (*b*) *Sclerophyton circumscriptum*, and (*c*) *Thrombium epigaeum* (after Lorrain Smith). Discomycetes: (*d*) *Scleroderris ribesia*, (*e*) *Helotium fructigenum*, (*f*) *Lecanidion atratum*, and (*g*) *Cryptodiscus pellidus* (after Boudier). *Right:* diagram of spores escaping from ascus; V_1 and V_2, volume below dotted line: p, hydrostatic pressure.

sharper apex and a blunter base, rarely if ever occurs. Since no developmental factors would appear to be responsible, it is reasonable to explore a 'teleological' explanation.

Bipolar asymmetrical ascospores are found only in Inoperculate Ascomycetes with successive discharge occurring through a small apical pore. As the ascospore is being pushed through the pore and is stretching it still wider, it may be moving quite slowly, and when the stretched pore is grasping the spore at its widest part, its velocity may be zero or nearly so. The work done thereafter by the hydrostatic pressure of the ascus is directed towards the actual discharge of the spore and not towards further stretching of the pore. This work is given by pV, where p is the hydrostatic pressure and V the volume of the part of the spore still immersed in the fluid of the ascus. Assuming the friction between spore and pore to be negligible, pV can be equated to $\frac{1}{2}mv^2$ where m is the mass of the spore and v its initial velocity of discharge. If this reasoning is justifiable, it is clear that for spores of equal mass, the nearer the widest part is to the front, the greater V and therefore, all else being equal, the greater the initial velocity, the greater the height of discharge, and the better the chance of effective dispersal.

In the fungi described so far the asci remain attached during discharge to a basal tissue within the perithecium, but in some flask-fungi the asci become detached at maturity. This seems to be particularly true of certain long-necked types. *Ceratostomella ampullasca* is an example. In this (Fig. 2.33) the perithecium is immersed in rotten oak wood with only the

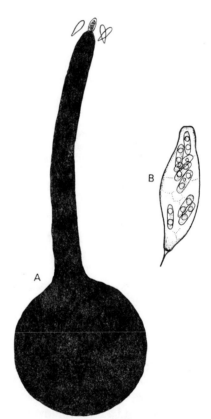

FIG. 2.33. *Ceratostomella ampullasca*. (A) perithecium mounted in water with an ascus at the ostiole about to discharge its spores. Three empty asci are seen near the ostiole and three groups of spores discharged from these are shown. (B) a single mature detached ascus. A × 65, B × 500.

tip of the long neck projecting. Within the perithecium the asci arise from a basal cushion but break free when ripe. The interior thus becomes filled with thousands of small asci, mostly detached. As they swell a pressure develops which is relieved by the asci passing out in single file through the narrow neck canal. In *C. ampullasca* the neck is opaque and their actual movement cannot be observed, but in *Gnomonia rubi*, due to the transparency of the neck, their passage has been watched (Fig. 2.34).

FIG. 2.34. *Gnomonia rubi*. Neck of the perithecium in optical longitudinal section showing passage of asci along neck-canal. Above the ostiole the spores and empty membrane of a discharged ascus are shown. ×450. After Dowson (1925).

In *Ceratostomella* there is a narrow ostiole through which the ascus squeezes until the spore-containing part is exposed. The ascus then bursts scattering its spores. Immediately afterwards the empty ascus is pushed out by the one below, and so on. This may be a quick-firing mechanism with only a few seconds elapsing between successive ejections. Had each ascus to elongate from a basal cushion up a neck canal 1–2 mm in length before discharge could occur, the rate of spore liberation would certainly be greatly reduced.

In perithecial Ascomycetes, although discharge is the rule, there are many examples where it does not occur. Further, certain species that normally discharge their spores violently may at times liberate them in a less active manner. As we have seen in *Daldinia concentrica* the ascospores are normally shot to a distance of 0·5–1·8 cm. However, under certain conditions discharge fails from certain perithecia of a stroma and a spore-tendril or 'cirrus' is produced at the ostiole. What happens is that the tip of an ascus reaches the ostiole, but instead of bursting explosively its contents merely ooze out. The next ascus behaves likewise and piles its eight spores just behind the first eight, pushing them forward a trifle. As the process goes on the cirrus grows longer and longer (Fig. 2.35). A similar behaviour of the asci seems to be the regular one in *Hypocrea pulvinata* in which no violent discharge occurs. This stromatal pyrenomycete is found commonly on the undersides of old sporophores of the bracket-fungus *Polyporus betulinus*. The ascospores collect on the surface of the stroma as a dryish powder fairly easily blown away by wind.

Non-violent ascospore liberation is also a feature of all species of the large genus *Chaetomium*, but the course of events leading to the escape of spores from the perithecium is somewhat different. Within the perithecium the ascus walls break down as the asci mature and the interior become filled largely with a mixture of free spores and mucilage which oozes out through the ostiole to form either a spore-tendril or a somewhat spherical mass (Fig. 2.36).

Closely related to *Ceratostomella ampullasca* are species of *Ceratocystis* in which violent discharge no longer occurs. The ascus walls disappear as in *Chaetomium* and the perithecial cavity is filled with a mass of spores and slime. This oozes up through the narrow neck-canal and at the ostiole a spore-drop is formed usually supported by a fringing ring of specialized supporting hyphae. The drop often becomes as large as, or larger than, the spherical part of the perithecium itself. The mucilage associated with the spores is usually rather viscous and the spore-mass frequently does not readily mix with water nor does it easily dry up. In some species, although the ascus walls break down, the tiny droplets, each of eight spores, do not lose their identity and, if a perithecium is mounted in water, instead of these droplets coalescing to form a single large drop they drift away individually. The long perithecium neck bearing its terminal mass of ascospores is an example of the stalked spore-drop (p. 67) which seems to have been evolved repeatedly in fungi. In *Ceratocystis* there is little doubt that the spores are insect dispersed.

Considerable attention has been given to the physiology of spore discharge in Pyrenomycetes. In the first place it must be emphasized that an over-riding factor in all Ascomycetes is the water supply, but consideration of this matter will be left to a later chapter when the water relations of spore liberation in fungi generally will be considered. At

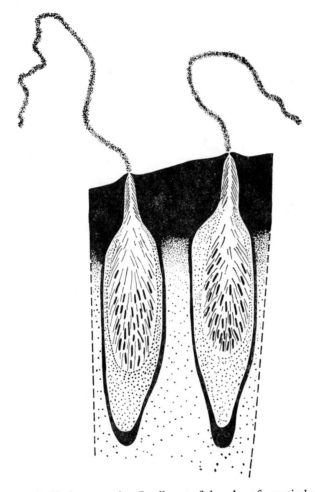

FIG. 2.35. *Daldinia concentrica*. Small part of the edge of a vertical section through the stroma showing two perithecia. Each short thick line in the perithecium represents a row of eight ascospores in an ascus. From both ostioles a 'spore-tendril' is emerging.

this stage the other major factors closely affecting discharge in flask-fungi will be discussed. These are light, the relative humidity of the ambient air, and temperature.

Much of our knowledge of the physiology of spore discharge relates to *Sordaria fimicola*. Light is an important factor (Ingold and Dring 1957). On a suitable medium this fungus fruits freely and discharges its spores both in darkness and in light. However, when a culture is subjected to a regime of 12 hours' light and 12 hours' darkness, with other conditions kept constant, discharge is mainly in the light periods (Fig. 2.37). At

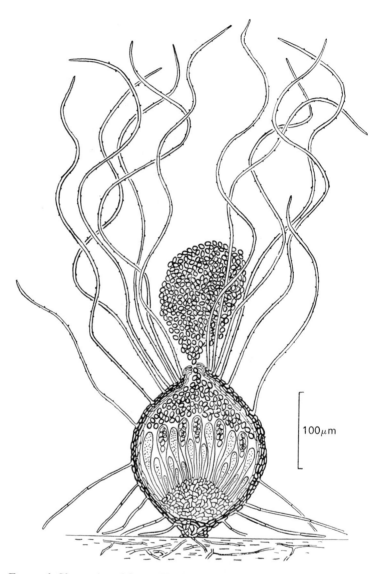

FIG. 2.36. *Chaetomium globosum*. Vertical section of a perithecium growing on agar. Asci break down within and the mass of spores oozes out through the ostiole.

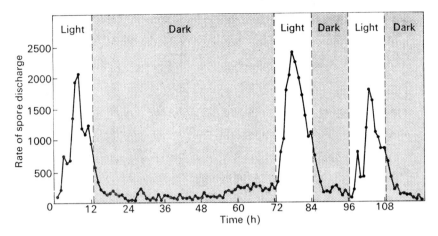

Fig. 2.37. *Sordaria fimicola*. Rate of spore discharge plotted against time for a culture which has, previous to the start of the experiment, been grown with 12 h light: 12 h darkness in each 24-h period.

each change from darkness to light discharge rises to a peak rate several hours later and then falls off.

Since transfer to light has such a profound influence on the rate of discharge, it is natural to consider how far very short periods of illumination may be effective. In a series of experiments involving the interruption of darkness with rather high light intensities (about 10 000 lx) operating for 10 minutes or less, it was found that 2–3 hours later the rate of discharge rose to a level of the order of ten times the original one, followed by a rapid decline (Fig. 2.38).

In the stimulation of discharge by light in *Sordaria* it is the blue end of the visible spectrum that is active, light of wavelength above 520 nm being ineffective (Ingold 1958). It is of interest to note that two other and very different processes, the phototropism of the perithecia and the production of a red carotenoid pigment, have essentially the same response spectra.

The effect of light on spore discharge has now been studied in a wide range of Pyrenomycetes. In sordariaceous fungi generally (especially in species of *Sordaria* and *Podospora*) light stimulates discharge. However, in *Apiosordaria verruculosa* there is a rather long interval (8–12 h) between the reception of the stimulus and maximum response. Xylariaceous fungi (e.g. *Hypoxylon*, *Xylaria*, and *Daldinia*) behave in the opposite manner with discharge directly inhibited by light.

There can thus be little doubt that under experimental conditions light is a major factor in spore liberation in flask-fungi, but it must never be forgotten that laboratory studies on the light factor have normally been

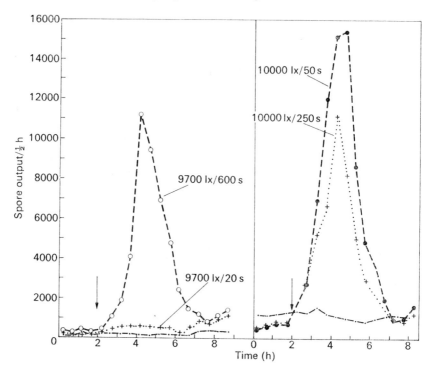

FIG. 2.38. *Sordaria fimicola.* Two sets of graphs showing rate of spore
discharge (in spores caught on 1 cm² surface above a discharging culture for
½ h) plotted against time. Prior to zero time cultures in dark for about 20 h.
Indicated short light treatment given at time shown by vertical arrow. In
each set of graphs there is an untreated control.

conducted under conditions where neither water supply nor temperature
are limiting.

The rate of spore discharge is, naturally, affected by temperature.
The effect of a particular temperature is, however, conditioned by previous
treatment. Thus in the dark, after 24 h at 20°C, transfer to 25°C soon led
to a great increase in the rate of discharge, but this was followed by a fall
and, at the end of 24 h at the higher temperature, the rate was little different
from that at 20°C. Further, a return to 20°C for the following day produced
little change in rate, but when the culture was again returned to 25°C
the peak was repeated (Fig. 2.39). When the same regime was used but
with alternating days of light and darkness in which the illuminated days
were either during the period of high or low temperature (Fig. 2.39), it
was found that spore discharge was at a much greater rate during the
days of light as compared with the dark ones, irrespective of temperature.
The familiar peak several hours after transfer to light is very evident.

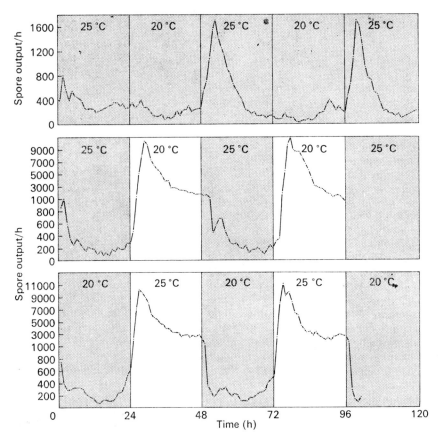

FIG. 2.39. *Sordaria fimicola.* Upper curve; rate of spore discharge from culture grown in the dark at 20°C and then subjected to alternate days at 25°C and 20°C in dark. Lower two curves: two parallel experiments on cultures reared in light at 20°C and then treated as shown with alternating days at 20°C and 25°C and alternating days of darkness and light (1000 lx, 'daylight' fluorescent) in one relative high temperature and darkness coincide; in the other relatively high temperature and light. In lower two curves note change of scale above 1000 spores per hour.

Essentially the same result was obtained when the contrasted temperatures were 15 and 25°C.

In continuous darkness with a regime involving days at 8°C alternating with others at 20°C, discharge fell quickly to nearly zero at the lower temperature and rose at the higher to a peak 8–10 h after transfer and thereafter rapidly declined. In a parallel experiment with cold periods illuminated, the picture was essentially the same (Fig. 2.40). Light appears to have no effect in stimulating discharge at 8°C. It thus seems that at higher temperatures light tends to be limiting, while when it is colder

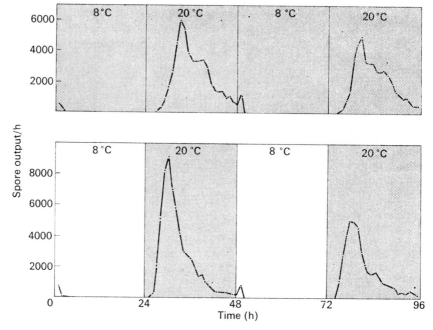

FIG. 2.40. *Sordaria fimicola.* Upper curve: culture reared in light at 20°C and then subjected in darkness to alternating days at 8°C and 20°C. Lower curve: parallel culture but illuminated (1000 lx) during cold periods.

temperature become the limiting factor. Under natural conditions, if water supply is adequate, both light and temperature probably co-operate to produce the day-time maximum of *Sordaria*.

Since in *Sordaria* only the lower part of the perithecium is submerged in the substratum (Fig. 2.20) while the neck projects, it might be expected that the humidity of the air around the ostiole itself could exert an influence on discharge. As this process involves the bursting of turgescent cells, it might further be thought that low humidity would militate against discharge. However, experiments show that low humidity, provided it does not persist too long, greatly stimulates discharge (Ingold and Marshall 1962). In an experiment, the results of which are illustrated in Fig. 2.41, a saturated air-stream and a very dry one were circulated over a culture in alternating hours. The rate of discharge was much higher in the dry as compared with the humid periods. However, with prolonged dry periods the rate, though initially high, steadily fell, and by contrast, in a following damp period, although discharge was at first at a very low level, it gradually rose. In a further study (Fig. 2.42) it has been found that the resulting increase in discharge following the substitution of dry for damp air is particularly apparent in the few minutes immediately following the

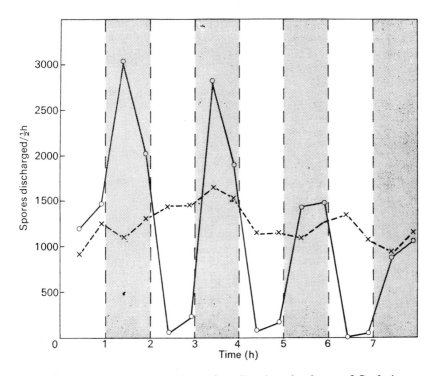

FIG. 2.41. Rate of spore discharge from illuminated cultures of *Sordaria* plotted against time. Dotted line: control culture in water-saturated air-stream. Continuous line: culture alternately in damp air-stream (un-stippled periods) and in dry (entering stream 35 per cent r.h.) air-stream (stippled periods). Both cultures before start of experiment subjected to damp air-stream for 15 h.

change. In explanation it has been suggested that incipient dessication of the perithecium may exert pressure on its contents so that the asci are brought more rapidly to the ostiole with a consequent increase in the rate of discharge (Austin 1968). A similar behaviour in *Claviceps purpurea* has been recorded by Hadley (1968). Few spores were discharged from the perithecial stroma in a saturated atmosphere, but lowering the relative humidity had a remarkable effect in increasing spore release.

Discharge in *Sordaria* is also influenced by increase of carbon dioxide in the air (Ingold and Marshall 1964). Concentrations as low as 0·2 per cent considerably stimulate discharge as compared with air from which the carbon dioxide has been removed. However, the amount of this gas normally present in the atmosphere (0·03 per cent) is without influence, air deprived of its carbon dioxide and normal untreated air having indistinguishable effects. Probably carbon dioxide has no significance as a factor influencing discharge in nature.

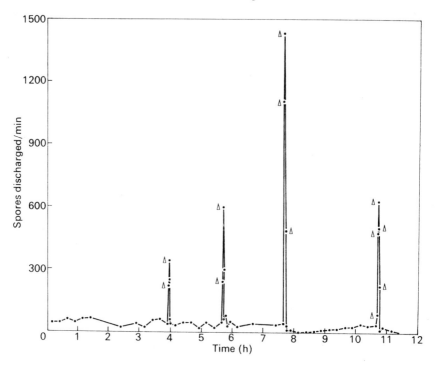

Fɪɢ. 2.42. Rate of spore discharge in *Sordaria fimicola* from a culture subjected to brief periods of 35 per cent relative humidity. Treatments were
for 2, 2, 3, and 5 min respectively, points marked with a triangle indicate
discharge under 35 per cent r.h. The culture was grown under constant
high humidity, and kept in an air-stream of 95 per cent r.h. for about 15 h
before the experiment. After Austin (1968a).

We are left with a picture of spore discharge in *Sordaria* as an active
process influenced by a range of external factors, but it is not clear at
what stage or stages in the development of an ascus a particular factor
exerts its effect. Temperature would presumably influence all enzyme-
controlled processes involved in ascus maturation, whereas light might,
perhaps, affect only one particular chemical reaction. Probably the effect
of humidity is physical rather than chemical.

Finally, it should be noted that there are internal as well as external
factors involved. Esser and Straub (1958) have made a study of the mutant
blocks in the maturation of *Sordaria macrospora* (Fig. 2.43). Each mutation
prevented further development beyond a certain point. One stopped the
final stage of actual discharge. Accepting the concept of 'one gene, one
enzyme', it would seem that some enzyme-controlled reaction is involved
in the final preparation for the bursting of the ascus. It would be of the
greatest interest in the present context to know the nature of this reaction.

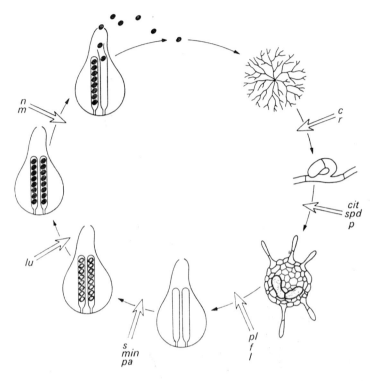

FIG. 2.43. *Sordaria macrospora.* Diagram showing the position of genetic blocks in perithecium development. The lettering indicates specific mutant genes which, when present in place of the normal alleles, block further development at the indicated positions in the process of perithecium and ascus formation. From Esser and Straub (1958).

Violent ascospore discharge is particularly associated with apothecia and perithecia, but it also occurs from the minute cleistothecia of powdery mildews. These are essentially hibernating structures that remain dormant throughout winter, but in spring show the renewed activity that eventually leads to the discharge of the spores. Within the firm wall of the cleistothecium are one or several oval asci. In some species only the ascospores are violently ejected; in others the asci are first shot away and then burst, scattering their spores.

Sphaerotheca mors-uvae is an example of the first type. The single ascus in the cleistothecium swells and this leads to rupture of the cleistothecium wall by a slit through which the ascus protrudes as it continues to swell. Finally the ascus bursts and the ascospores are squirted out (Fig. 2.44).

In *Podosphaera leucotricha* double discharge occurs. This has been observed by Woodward (1927) and is best described in his own words:

'The perithecium is a true cleistocarp, the ascus being ejected though an irregular rupture of the wall. The opening formed gradually widens owing to the expansion of the ascus following the absorption of water . . . The perithecial wall is elastic and responds to the pressure exerted from within until the opening, although small in diameter, can no longer retain the ascus which squeezes through the orifice and is violently ejected. The force of ejection is considerably increased by the snapping together of the "jaws" of the orifice, due to the elasticity of the perithecial case . . . The ascus is often thrown several centimetres in the air. On reaching water or in a damp atmosphere, it continues to swell and within one or two minutes explodes, scattering the ascospores in all directions.'

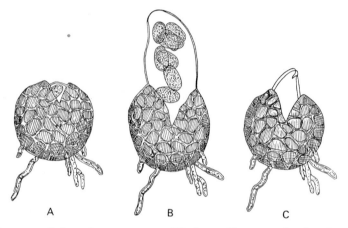

FIG. 2.44. *Sphaerotheca mors-uvae.* (A) the swelling ascus has just burst the wall of the cleistothecium. (B) the ascus is fully swollen and about to discharge. (C) just after discharge. After Salmon (1914).

From the point of view of dispersal the monotypic genus *Phyllactinia* deserves special consideration (Fig. 2.45). *P. corylea* is found in Britain on leaves of hazel. The cleistothecium is a remarkable structure. First each appendage is a unicellular thorn-like structure with a bulbous swelling where it is attached. The appendage is thick-walled, but not uniformly so, for the lower half of the bulbous part has a thin wall so that, on drying, the appendages change shape, with the result that the cleistothecium is raised from the leaf surface, apparently preparing it for take-off. The second curious feature is the presence on top of a tuft of compound hairs that secrete a relatively large drop of viscous slime. It is difficult to resist the view that the cleistothecium of *Phyllactinia* is itself a dispersal unit, although at a later stage ascospore liberation no doubt occurs as in other powdery mildews. The functioning of this elaborate apparatus in the

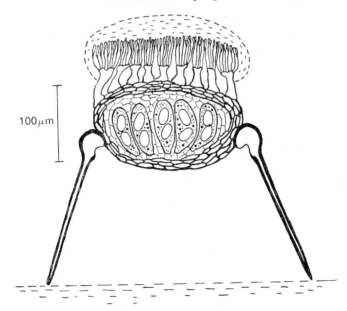

FIG. 2.45. *Phyllactinia corylea*. Vertical section of cleistothecium in the 'raised' position. Each ascus contains two ascospores.

dispersal story of the fungus is far from clear and a critical study of its biology is long overdue.

Spore discharge in *Protomyces* may be considered at this point because it resembles, somewhat, discharge in *Sphaerotheca*. *Protomyces* is one of those difficult genera that have failed to settle neatly into any scheme of classification. Some have placed it in Phycomycetes as an archimycete type; others have tried to push it into Ustilaginales; others, in despair, have made for it a special group of fungi, Protomycetes; but now most mycologists are resigned to including it amongst the 'lower' Ascomycetes.

Species of *Protomyces* occur as obligate parasites on herbaceous plants, but are of no economic importance. *P. macrosporus* on *Aegopodium*, *P. pachydermus* on *Taraxacum*, and *P. inundatus* on *Apium nodiflorum* are common species. They produce small warts on the stalks and veins of the leaves and on the stems. Each wart is the result of a single infection, and, in a section of a wart, the fungus is seen as an intercellular mycelium of limited extent with many much-enlarged intercalary 'chlamydospores' each with a two-layered wall. These are resting spores in *P. macrosporus* and *P. pachydermus*, but in *P. inundatus* they germinate at once. When fully developed, the warts on *Apium* split open exposing the parasite. Each exposed 'chlamydospore' may be regarded as an ascus confined within an outer rigid cell-wall. As the cell absorbs water the ascus expands, bursts through the outer rigid cell-wall and lies free as a spherical cell containing a large number of minute spores clumped together. The 'ascus' then bursts by a slit and the spores are thrown into the air to a distance of about 2 cm (Fig. 2.46).

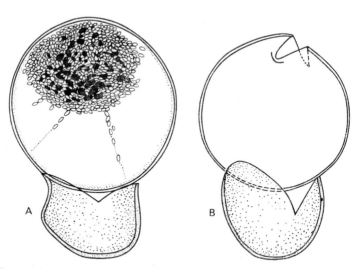

FIG. 2.46. *Protomyces inundatus*. (A) just before discharge; the ascus has burst free from its outer thick wall and has swollen considerably. (B) after discharge. × 320.

3. SPORE LIBERATION IN MUCORALES

IT is proposed in this chapter to consider spore liberation in Mucorales, essentially a 'natural' group of terrestrial fungi which, in spite of its small size (about 45 genera and 275 species) shows a wide variety in its asexual apparatus related to dispersal.

In Mucorales the mycelium is well developed and without septa when young. Asexual reproduction is typically by non-motile spores produced in sporangia, or if conidia are formed instead of, or in addition to, sporangia, they can reasonably be interpreted as the ultimate stage in sporangial reduction. The asexual spores are normally raised above the substratum on erect sporangiophores or conidiophores. Further, the group is characterized by a distinctive sexual process resulting in zygospore formation.

The general biology of Mucorales is interesting. Many are soil saprophytes, especially species of *Mucor*, *Mortierella*, *Absidia*, *Cunninghamella*, and *Syncephalastrum*. Some occur with great regularity on the dung of larger herbivores. Thus in the early phases of the succession on horse-dung species of *Mucor*, *Pilaria*, and especially *Pilobolus* normally appear. Some members of the order develop on the droppings of rodents, including saprophytes belonging to the genera *Mucor*, *Phycomyces*, *Thamnidium*, *Helicostylum*, *Coemansia*, *Kickxella*, *Mortierella*, and *Haplosporangium*. Both on horse- and on rodent-dung the saprophytic species are normally attacked at a later stage by parasitic members of the same order distributed among a number of genera: *Chaetocladium*, *Piptocephalis*, *Syncephalis*, *Dispira*, and *Dimargaris*. However, in these coprophilous Mucorales the distinction between parasites and saprophytes is not always easy to draw. In most species of *Piptocephalis* parasitism seems to be practically obligate, but others, for example the common *Chaetocladium brefeldii*, grow and sporulate readily on nutrient agar, but are probably always parasitic in nature.

Both the saprophytic and the parasitic Mucorales of dung appear to have the same dispersal story. The spores are eaten incidentally with food and, having passed through the alimentary canal, germinate subsequently in the excrement. It is interesting to note how varied is the mucoraceous flora of the droppings of rodents especially field mice, and it is difficult to resist the impression that members of the order reach their fullest expression on this unlikely substratum. The mucoraceous flora on mouse dung is distinctly richer in species than that on horse dung.

However, species of *Pilobolus*, such a striking feature of the early coprophilous succession on horse dung, seem to be rare on the dung of field mice.

Another biological group in Mucorales consists of species found on the larger fungi. *Syzygites megalocarpus* (= *Sporodinia grandis*) is a common mould on a wide range of decaying agarics. The much rarer *Dicranophora fulva* has been found only on *Paxillus*, *Gomphidius*, and *Boletus*, now generally regarded as closely related toadstool genera. Species of *Spinellus* are common on *Mycena* and *Collybia* and are probably parasitic.

Some Mucorales are unspecialized parasites of the sappy tissues of higher plants. Thus *Rhizopus stolonifer* commonly causes a soft-rot of fruit such as apples, plums, and tomatoes; in the warmer countries *Choanephora cucurbitarum* brings about decay in the flowers and fruit of cucurbits; and *Gilbertella persicaria* is responsible for a dry-rot in peaches.

Finally, rather sharply separated from other Mucorales, are the hypogeal fungi grouped in Endogonaceae, a family closely parallel to Tuberales in Ascomycetes and to Hymenogastrales in Basidiomycetes. In *Endogone*, for example, the reproductive bodies (sporangia, zygospores, or chlamydospores) are usually aggregated into small but macroscopic sporophores and, as with truffles, dispersal is presumably by rodents that grub up and eat these fruit-bodies. Indeed, in North America the characteristic chlamydospores of *Endogone* have been found abundantly in the intestinal tracts of deer mice (Dowding 1955). The Endogonaceae will not be considered further in this chapter.

Benjamin (1959) recognizes eleven families in Mucorales, but their interrelationships are more than usually dubious. However, from the functional rather than the systematic point of view, four fairly distinct types of asexual apparatus seem to be involved (Ingold and Zoberi 1963). The best known is the large spherical or sub-spherical sporangium with a substantial columella and containing rarely less than a hundred, and usually many thousand, spores. In the relatively huge sporangia of *Phycomyces blakesleeanus* the number may reach 100 000. Secondly, there is the sporangiole; a small nearly spherical sporangium with from two to about twenty spores and with the columella usually poorly developed. A third type is the merosporangium; really a cylindrical sporangiole containing from two to a dozen spores in a single row. Finally there are conidia. In certain species it is clear that they are to be interpreted as reduced sporangioles or reduced merosporangia. However, in others (e.g. species of *Cunninghamella* and *Mycotypha*) the validity of this interpretation is by no means clear and it seems best, in general, to use the non-committal term 'conidium'. These four types will be considered in some detail.

In species with relatively large columellate sporangia three general arrangements can easily be recognized: those in which direct liberation into the air is unlikely because the spores are associated with sticky mucilage; those in which the spores form a dry friable mass on dehiscence of the sporangia so that they are readily blown away; and species of *Pilobolus* in which the whole sporangium is violently discharged.

Among the first group are some species where the spores are suspended in a mucilaginous matrix within a very thin sporangium wall which, nevertheless, persists until it is ruptured by some external influence. To this subdivision *Phycomyces*, *Spinellus*, and some of the larger species of *Mucor* (e.g. *M. plasmaticus*, Fig. 3.1D) belong. Of special interest is *Phycomyces*, the commonest species being *P. nitens* (Fig. 3.1A, B).

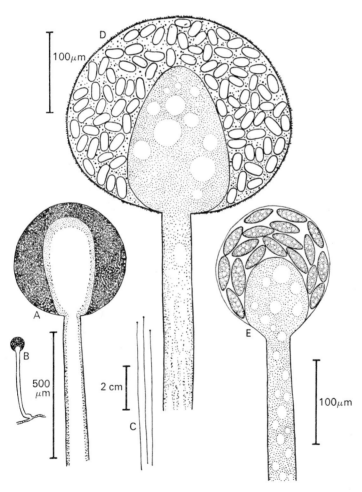

FIG. 3.1. A–C, *Phycomyces blakesleeanus*: (A) large sporangium; (B) dwarf sporangiophore at same magnification; (C) three sporangiophores. (D) *Mucor plasmaticus*, (E) *Spinellus fusiger*.

This has been isolated on many occasions, nearly always from the droppings of rats and mice. The steel-grey sporangiophores are unbranched, strongly phototropic, and may reach a height of 20 cm.

A remarkable feature of *Phycomyces* may be noted, namely its ability to form its sporangiophores even when the humidity of the air is maintained at a very low level. If specimen tubes ($2 \cdot 5 \times 2 \cdot 5$ cm) brimful with malt agar are inoculated with *Phycomyes* and some are placed in a dessicator over anhydrous calcium chloride while others are over water, the sporangiophores are produced equally freely under both conditions, whereas parallel cultures of *Mucor*, for example, produce no aerial growth in the dry air. This ability of *Phycomyces* seems to be due to the impermeable nature of the mature wall of the sporangiophore practically restricting evaporation to the narrow growing zone just below the sporangium, and to the well-developed transpiration stream making good such water as is lost. If an opened Petri-dish culture is examined under the low power of the microscope, the rapidly flowing contents in the principal hyphae are easily seen, and if a particular stream is followed it always leads into the base of a sporangiophore. This streaming depends on transpiration and the hyphal contents flow to the site of water loss. With such a striking ability to erect long aerial sporangiophores in dry air, it is remarkable that the sporangium does not liberate dry spores. Had this ability been coupled with a sporangial apparatus like that of *Rhizopus*, *Phycomyces* might well be much more abundant.

As pointed out by Dobbs (1939), the sporangium in a number of the commonest species of *Mucor* (e.g. *M. hiemalis* and *M. ramannianus*) is converted at maturity into a sporangial drop. The sporangium wall seems to dissolve except for a basal region which persists as a minute collar round the columella (Fig. 3.2C, D). Ordinarily, in Petri-dish cultures the sporangial drop is several times the volume of the original sporangium. Under conditions of low humidity the drop, which apparently contains mucilage, dries leaving the spores firmly cemented to one another and to the columella. Strong air currents or vigorous tapping of inverted cultures fail to dislodge the spores. This kind of sporangial behaviour is also seen in *Dicranophora fulva* and in the terminal sporangia of *Thamnidium elegans* and *Helicostylum fresenii*. A slight modification is to be found in *Gilbertella persicaria* (Fig. 3.2E, F) a member of the Choanephoraceae. The sporangium wall is persistent, but dehiscence is by a line of weakness running from pole to pole around the spherical sporangium. It seems that, in addition to the spores, the sporangium contains mucilage which swells pushing apart the two halves of the wall.

Pilaira anomala, a common coprophilous species, also produces a sporangial drop (Fig. 3.2A, B). In *Pilaira* and *Pilobolus*, the only two genera of the family Pilobolaceae, the sporangium is distinctive. Its wall is differentiated into an upper resistant blackened part and a lower

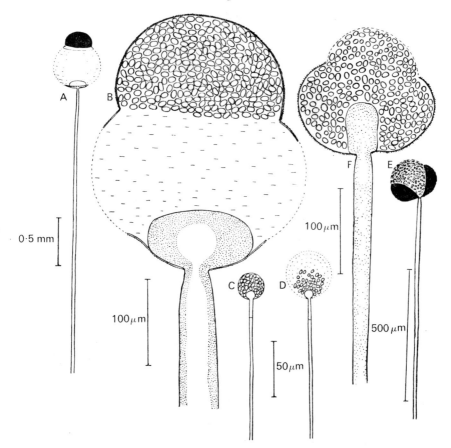

Fig. 3.2. A–B, *Pilaria anomala*: (A) upper part of sporangiophore with dehisced sporangium now forming a sporangial drop; (B) similar drop in vertical optical section. C–D, *Mucor ramannianus*: (C) undehisced sporangium; (D) sporangium now converted into a sporangial drop. E–F, *Gilbertella persicaria*: (E) sporangiophore with dehisced sporangium and sporangial drop exuding; (F) similar sporangium in vertical optical section.

delicate semi-transparent part which is white in reflected light. Within are spores and mucilage, the spores being concentrated in the upper part. In both genera the mature sporangium ruptures in the delicate region of the wall around a circle of latitude well below the equator of the sporangium. After sporangial dehiscence the mucilage swells considerably and is directly exposed in the gap created by dehiscence. Thus a sporangial drop is formed and this may be carried onto the herbage around the dung by the growth of the unbranched phototropic sporangiophore which elongates from a few millimetres to several centimetres in a few hours. When the delicate sporangiophores wither, the sporangia, without their columellas, are left adhering to the grass just as if they had been discharged there (Fig. 3.3).

The sporangial drop is an example of the 'stalked spore-drop', a distinctive type of spore-presentation mechanism probably associated most often with dispersal by insects or rain-splash. As will appear later, it is found in some merosporangiferous genera of Mucorales and in

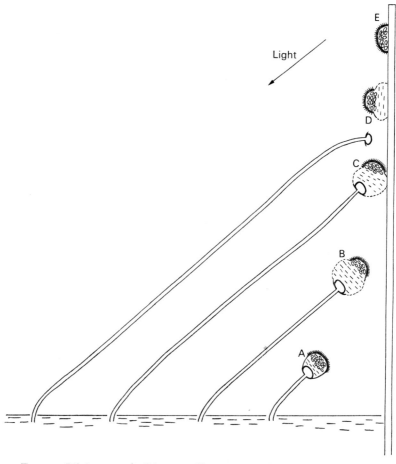

Fig. 3.3. *Pilaira anomala*. Diagram of how a sporangium, carried to a neighbouring object (here a vertical sheet of glass) by elongation of the sporangiophore, becomes attached so that the black cap of the sporangium covers the spores. (A) phototropic (but short) sporangiophore oriented towards light; sporangium undehisced. (B) elongating sporangiophore with sporangium now dehisced and mucilage greatly swollen to form sporangial drop. (C) drop just touching glass, (D) drop stuck to glass and sporangiophore beginning to wither; (E) sporangiophore has disappeared and mucilage of sporangium has dried. Approx. × 20; but sporangiophores in C and D are shown much shorter than they actually are.

Kickxellaceae. Further, outside the order this kind of apparatus is encountered repeatedly. Figure 3.4 illustrates some examples. In *Dipodascus* the stalk is an empty ascus and the drop consists of the escaped ascospores and sap. In *Cephaloascus* the stalk is a diploid ascophore and the drop is a mass of spores derived from the deliquescence of a head of asci. In *Ceratocystis* the stalk is the neck of a perithecium and the drop

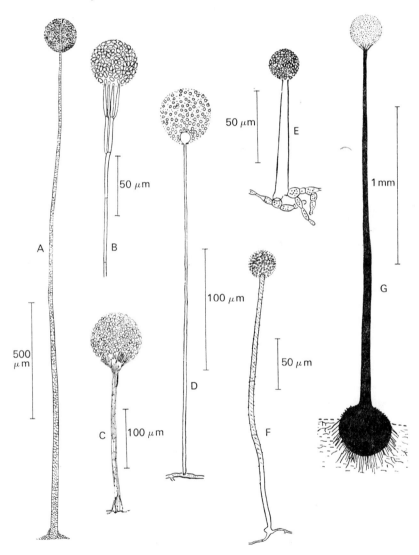

FIG. 3.4. (A) *Dictyostelium discoideum;* (B) *Gliocladium roseum;* (C) *Graphium cuneiferum;* (D) *Mucor ramannianus;* (E) *Dipodascus uninucleatus;* (F) *Cephaloascus fragrans;* (G) *Ceratocystis adiposa.* All, except *Ceratocystis* shown in longitudinal section.

consists of extruded ascospores. In *Gliocladium* the conidiophore is of the *Penicillium* type but the phialospores, being produced with slime, form a drop, and *Graphium* is similar but the stalk is a sheaf of parallel hyphae. In *Dictyostelium*, probably not placeable in the Fungi at all, the stalk is a cellulose tube stuffed with inflated encysted amoebae, and the drop is composed of spores in a fluid matrix.

It certain species, not only of *Mucor* but also of some closely related genera (especially *Rhizopus*, *Actinomucor*, and *Circinella*), the spores are not associated with mucilage in the sporangium; dehiscence by rupture of the wall takes place and direct aerial dispersal is possible. A good example is *M. petrinsularis* (Fig. 3.5B, C). The ripe intact sporangium is packed with spores which, being closely confined between the sporangium wall and the columella, are compressed into polyhedral shapes, a condition not common in mucoraceous sporangia. As the spores mature their tendency to round off increases and finally the sporangium wall ruptures in a somewhat irregular manner to form a kind of saucer with most of the spores lying on its upper surface, although a few adhere to the columella. The wide opening of the sporangium, a special feature of this species, seems to be due to swelling of those spores which during development occupy the somewhat acute angle between the base of the columella and the sporangial wall. Dehiscence of the sporangium occurs under conditions of saturation and is not related to drying. If a young sporulating culture on nutrient agar in a closed Petri dish is inverted and tapped, many spores drop on to the inner surface of the lid, mostly in groups attached to segments of sporangium wall.

The behaviour of the common species *M. plumbeus* is somewhat similar (Fig. 3.5D). Under dry conditions the sporangium wall breaks into a few separate, angular fragments. Spores are readily blown away by wind. Usually each wall-fragment carries a small load of spores, although some spores escape singly.

Actinomucor elegans, a frequent species of apparently world-wide distribution, behaves in a somewhat different manner (Fig. 3.5A). This fungus is isolated from *Mucor* in a separate genus by virtue of its stoloniferous habit. In *A. elegans* the whole spore-mass enclosed by the sporangial wall slips off the smallish columella and is dispersed as a whole. This release can be brought about either by wind or by mechanical agitation.

In some other mucoraceous moulds dry spores are liberated, mostly singly, when the sporangia dehisce on drying. A well-known example is *Rhizopus stolonifer* (Fig. 3.6). The mature sporangium has a very large columella. On exposure to dry air this collapses in a striking and very definite manner so that it looks like an inverted pudding-bowl balanced on the end of a stick represented by the relatively short, stiff sporangiophore. Apparently at the same time, and possibly as a consequence

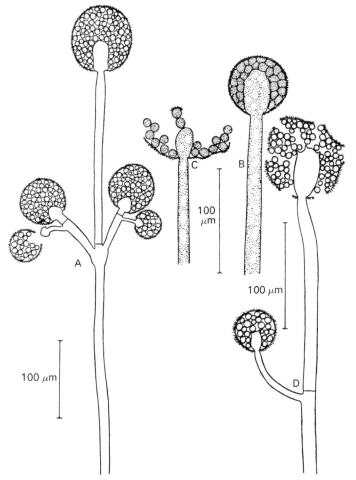

Fig. 3.5. (A) *Actinomucor elegans:* part of sporangiophore; one lateral sporangium has just separated from its columella. (B–C) *Mucor petrinsularis;* intact sporangium and another after dehiscence under damp conditions. (D) *M. plumbeus:* apical region of sporangiophore, the terminal sporangium has dehisced, the lateral one is still intact.

of columella collapse, the brittle sporangial wall breaks into many small angular fragments. These with the dry spores form a powder easily blown away.

As in *Rhizopus stolonifer* so in *Circinella umbellata* (Fig. 3.7C–E), and probably in other species of that genus, spores are shed dry and mostly singly. The sporangial wall breaks down on drying along a few irregular radial fractures exposing the spores which sift away. A dry-spore mechanism also occurs in *Syzygites megalocarpus* (Fig. 3.7A, B). The sporangiophore, which may be several centimetres long, branches dichotomously

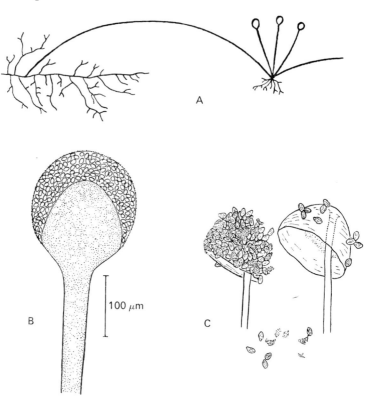

FIG. 3.6. *Rhizopus stolonifer.* (A) an aerial stolon from behind the growing margin of a culture on agar has made contact with the medium again and produced a group of three sporangiophores and a system of rhizoidal hyphae; (B) a sporangium in longitudinal section; (C) Two sporangia which have dried and their columellas have collapsed; in one the load of dry spores and wall fragments is still intact; in the other they have mostly been blown away.

and repeatedly in the apical region, each ultimate ramulus bearing a small sporangium with a relatively large columella and a few rather big spores. At maturity the very thin sporangial wall completely disappears and the spores are left exposed on the columella from which they are easily blown away.

It might be observed at this stage that, in spite of their abundance, mucoraceous moulds do not figure prominently in the air spora. An extensive aeromycological study has been made in Kansas, U.S.A. (Kramer, Rogerson, and Ouye 1959) lasting over two years and during this period Petri dishes with nutrient agar were exposed almost daily in an electrostatic bacterial air sampler, the developing colonies being subsequently identified to the genus. Of 113 667 colonies obtained, only

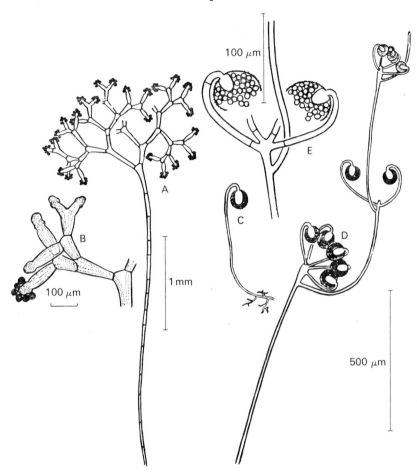

FIG. 3.7. A–B, *Syzygites megalocarpus*: (A) sporangiophore with ripe sporangia, only part of the unbranched stalk shown; (B) part of branch system, one columella still has its layer of spores. C–E, *Circinella umbellata*: (C) unbranched sporangiophore arising from feeding mycelium; (D) usual type of sporangiophore with groups of sporangia and sympodial development; (E) details of a sporangial group showing two dehisced sporangia.

194 belonged to Mucorales and of these 156 were of *Rhizopus*. By way of contrast the figure for *Cladosporium* was 50 548.

Pilobolus (Plate III) stands on its own as the only genus in Mucorales exhibiting violent spore discharge. It has attracted attention since it was first figured and crudely described towards the end of the seventeenth century. The history of research on *Pilobolus* and its general biology was fully summarized by Buller in the sixth volume of his *Researches on fungi* in 1933, and since that time it has continued to attract research workers particularly in relation to its specialized nutrition, its photo-tropism and the periodicity of its sporangial discharge. This last subject will be discussed in a later chapter, and it is sufficient here to note its diurnal rhythm of ripening and discharge under natural conditions.

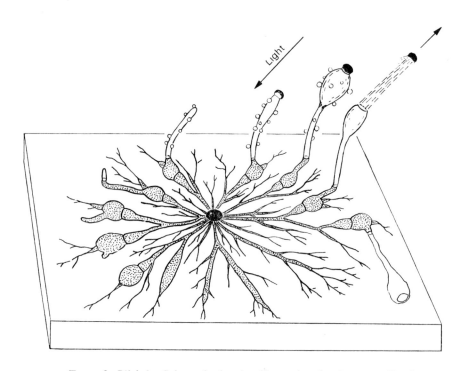

FIG. 3.8. *Pilobolus* Schematic drawing illustrating development. Tropho-cysts and sporangia are represented as developing in clockwise sequence (starting at 4 o'clock) on hyphae emanating from one or more spores of a sporangium planted on agar. The parts at and below the surface of the agar are stippled; parts above the agar are unstippled. Slightly modified after Page (1962).

Pilobolus is a genus of about half a dozen species several of which are common. On horse dung the most abundant is *P. kleinii*, but *P. sphaero-sporus* seems to be an easier species to grow and maintain in pure culture in the laboratory. The developmental story of *Pilobolus* is illustrated in Fig. 3.8. From an intercalary carotene-rich 'trophocyst', delimited by cross-walls from the feeding mycelium, the sporangiophore arises as an erect phototropic aerial hypha. From the apex of this a sporangium is produced which is ultimately cut off by a highly arched cross-wall forming a substantial columella. Later the region of the sporangiophore immediately below swells to form the subsporangial bulb. The mature sporangiophore (Fig. 3.9), including its parent trophocyst, is a single turgid cell. The stretched cell-wall has a lining layer of protoplasm. This layer is considerably thicker at the base of the subsporangial bulb and in most species this region is rich in oil droplets containing carotene giving it a conspicuous orange colour particularly in *P. kleinii*. The large vacuole is occupied by cell sap with an osmotic pressure of about seven atmospheres (Buller 1933).

The ripe sporangium undergoes dehiscence along a subequatorial line leaving a circular gap of exposed mucilage, and shortly afterwards discharge takes place. This involves the rupture of the subsporangial bulb just below the junction with the columella. Immediately this occurs the stretched wall of the sporangiophore contracts, squirting out a jet of sap that carries the sporangium together with the columella to a distance of up to 2·5 metres.

Using a set-up involving the triggering of a camera by a photoelectric cell activated by the disappearance of the black sporangium from the field of view of a microscope, Page (1964) managed to photograph the issuing jet capped by the sporangium at the precise moment of discharge. Initially this is a slightly tapering cylinder (Fig. 3.8), but in a split second it breaks up into a major apical drop, associated with the sporangium, followed by a tail of much smaller droplets.

Much ingenuity has been lavished on determining the velocity of the sporangium just after discharge. Pringsheim and Czurda (1927) used a set-up consisting of two horizontal disks, 30 cm in diameter and 10 cm apart, rotating rapidly on the same vertical axis with the lower disk 2·5 cm above a discharging culture of *Pilobolus*. This disk had a radial slot and any sporangium that happened to pass through it was impacted on the second disk. The horizontal displacement of the sporangium on the upper disk, relative to the slot on the lower disk, could be measured. Then knowing the speed of rotation of the disks, the velocity of projection during the early part (from 2·5 to 12·5 cm) of the flight could be determined. A mean value of 14 m/s was obtained.

Page and Kennedy (1964) employed other methods. In the first a photocell arrangement was used to time the minute interval between the interruption of two beams of light (3 mm apart) by a passing sporangium just after take-off. With this method values were obtained ranging from 4 to 15 m/s.

In the second method a block of agar with mature sporangiophores was set on a thin rigid membrane (fixed to a stiff wire) which acted as a minute launching platform. This vibrated when a sporangiophore burst

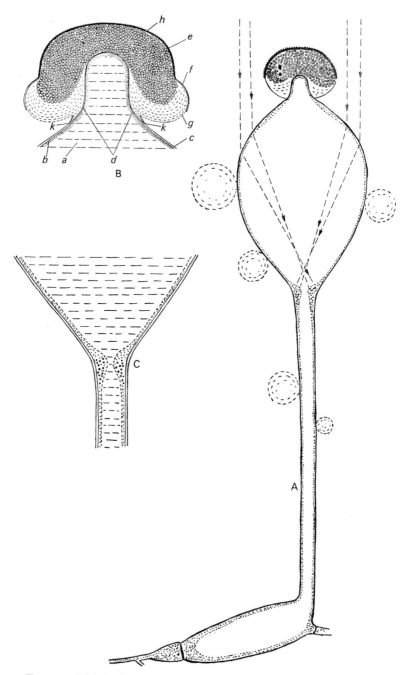

Fig. 3.9. *Pilobolus longipes*. (A) L.S. sporangiophore at a stage when the sporangium has not yet dehisced. Liquid droplets are exuded from the sporangiophore. Dotted lines show how incident light is focused on the 'retina'. (B) L.S. sporangium just before discharge: (*a*) sporangiophore sap; (*b*) layer of protoplasm; (*c*) sporangiophore wall; (*d*) columella; (*e*) opaque black sporangial wall; (*f*) semi-transparent part of wall; (*g*) mucilage; (*h*) spores; (*k*) region where sporangiophore ruptures. (C) base of subsporangial bulb, note thickened, carotene-rich protoplasm forming 'retina'.

and a second similar membrane, 1·5 cm from the first, vibrated when struck by the sporangium. Using an ingenious set-up involving phonograph cartridges, a tape recorder, and a dual-beam oscilloscope, they were able to measure precisely the time interval between discharge and impact. Extensive results were obtained with this method, the values ranging from 4·7 to 27·5 m/s.

Page and Kennedy studied the relation between the size of the sporangium and the initial velocity of its discharge. The latter tended to increase with size but the correlation coefficient, though positive, was low (about 0·2).

The columella and the mucilage of the sporangium are freely wettable but the upper blackened part of the sporangial wall, studded with minute projecting crystals of calcium oxalate as in most species of *Mucor*, is unwettable and thus projects from the drop of sap. Further, when the drop strikes an object, although the sporangium may momentarily be submerged, it bobs up again exposing the unwettable part. As a result, when the sap and the mucilage dry, the spore-mass, roofed over by the black resistant part of the sporangium wall, is completely protected and sticks fast (Fig. 3.10). Discharged sporangia may remain thus for many days before the grass is eaten by a herbivore and the dispersal cycle completed by passage through the animal.

In the water-squirting gun of *Pilobolus* the spore projectile is situated on top of the bursting cell, not within as with the ascus. There are a few examples elsewhere in the Fungi of similar mechanisms, and these may be briefly considered at this stage. Two are to be found in Entomophthorales, usually classified close to Mucorales, and the third occurs amongst 'imperfect' fungi.

Particularly comparable with *Pilobolus* is *Basidiobolus*, well known from the type species, *B. ranarum* (Fig. 3.11). This occurs with considerable regularity on the excrement of frogs, the conidiophores projecting into the air like minute specimens of *Pilobolus* (Plate III). The conidiophore, which is positively phototropic, arises from a single cell of the septate mycelium in the excrement. There is a straight stalk and a sub-conidial bulb the tip of which projects as a minute columella into the spore. The conidiophore is a highly turgid cell that finally bursts around a transverse circle of weakness near the base of the bulb, not near the top as in *Pilobolus*, and the elastic part of the stretched wall is above this line, not below it. Thus, on bursting, the stretched wall of the upper part contracts, squirts sap backwards, and flies off on the recoil carrying the conidium with it. It is a minute rocket (Ingold 1934). During the flight through the air the two parts of the projectile usually separate. Although the conidium is shot to a distance of 1–2 cm, the accompanying portion of the conidiophore (the 'basidium') is rarely thrown further than 0·5 cm.

Another example is *Entomophthora muscae*. House flies are frequently attacked by this parasite in late summer. Just before death a doomed

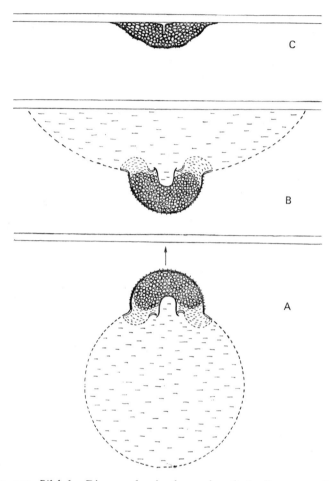

Fig. 3.10. *Pilobolus*. Diagram showing how, when the leading sporangium-bearing drop (A) of the sporangiophore sap-jet strikes an object, the sporangium with its unwettable black wall comes to the surface (B) and on drying adheres firmly (C) with the spore-mass covered by the black resistant part of the sporangium wall. Considerably modified after figures by Buller (1934).

fly, its abdomen visibly swollen with the fungus, attaches itself to some object, often a pane of glass, by its sucker-like proboscis. Then in a few hours thousands of unbranched conidiophores protrude between the abdominal segments forming several white transverse bands. A conidium is borne on each conidiophore, which is somewhat inflated, and this finally bursts shooting the conidium to 1–3 cm. The dead insect becomes surrounded by discharged conidia forming a white halo several centimetres across (Fig. 31.2).

In *Nigrospora sphaerica*, a member of Fungi Imperfecti occurring as a saprophyte on stalks of dead grasses (e.g. *Dactylis*), the penultimate cell of the short stout conidiophore is extended as a fine tapered process through the ultimate cell so that it almost makes contact with the base of the large (20 μm diam) black conidium. It is, apparently, by the bursting of this penultimate cell that the jet of fluid is produced that carries the

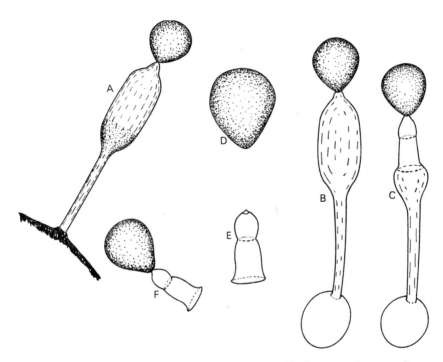

FIG. 3.11. *Basidiobolus ranarum*. (A) mature condiophore growing out of frog's excrement; (B) living conidiophore derived from a discharged conidium and mounted in water; (C) the same after addition of trace of iodine; (D) discharged conidium; (E) discharged 'basidium'; (F) a conidium with 'basidium' that has failed to separate. A, B, and C ×500; D, E, and F ×600.

conidium to a distance of 2–6 cm, the pierced ultimate cell being left behind on the conidiophore (Webster 1952). It is no accident that *Nigrospora* bears a certain resemblance to *Pilobolus* (Fig. 3.13). The discharge mechanisms are essentially similar.

Returning to Mucorales, it is quite clear that *Pilaira* and *Pilobolus* resemble one another much more than either resembles any other genus, but there remains the fundamental difference that in one violent discharge occurs, while in the other it does not.

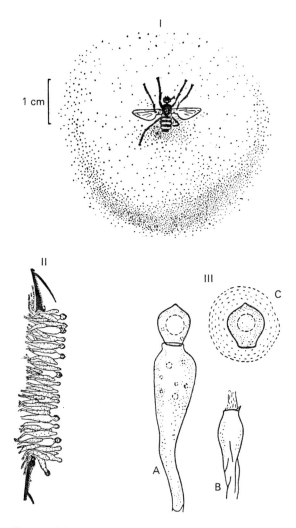

FIG. 3.12. *Entomophthora muscae*. (I) Dead fly attached to glass by its mouth-parts and surrounded by a halo of discharged conidia. Note the three white bands of conidiophores on its abdomen. (II) Part of a section through the abdomen of a dead fly showing the conidiophores projecting through the integument of the insect. × 100. (III) (A) single mature conidiophore; (B) conidiophore after discharge; (C) discharged conidium surrounded by a drop of conidiophore sap. × 400.

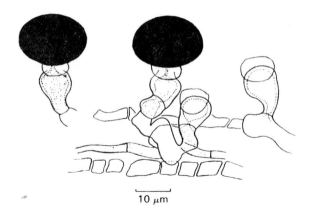

10 μm

FIG. 3.13. *Nigrospora sphaerica*. Two conidiophores with mature spores. Below each black conidium is a small 'supporting cell' and below this the 'ampulliform cell' which supplies the liquid jet for discharge. On the left are two conidiophores from which spores have been discharged. After Webster (1952).

Having considered spore-liberation in those Mucorales with large columellate sporangia, we may now pass on to consider the forms producing sporangioles. These are included in Thamnidiaceae and Choanephoraceae. In some, for example *Thamnidium elegans*, *Helicostylum fresenii* (Fig. 3.14), and *Choanephora trispora* (Fig. 3.16), sporangia as well as sporangioles occurs, in others such as *Cokeromyces recurvatus* and *Radiomyces spectabilis* (Fig. 3.15) only sporangioles are produced. In all these fungi the sporangioles are deciduous and are easily blown off by wind. The sporangiole bears with it a length of stalk which is usually very short, but is relatively long in *Helicostylum piriforme* (Fig. 3.14G). Most sporangioles are indehiscent. However, in *Choanephora trispora* the liberated sporangiole dehisces if it falls into water, its wall splitting into two hemispherical shells to liberate the spores with their polar bristles extended (Fig. 3.16C). Again in *Radiomyces spectabilis* the sporangiole breaks down in water setting free its spores.

It should be emphasized that, although the sporangioles of *Thamnidium* and *Helicostylum* are deciduous, the sporangia behave in a different manner, remaining attached and becoming converted into sporangial drops as in *Mucor hiemalis*.

The merosporangiferous Mucorales have been reviewed in an outstanding monograph (Benjamin 1959). Merosporangia are a feature of three familes: Piptocephalidaceae, Syncephalastraceae, and Dimargaritaceae. In all merosporangia the sporangial wall breaks down at maturity so that a row of spores, sometimes called a 'spore-rod', is formed looking superficially like a chain of conidia; the chain being reduced to only two spores in Dimargaritaceae.

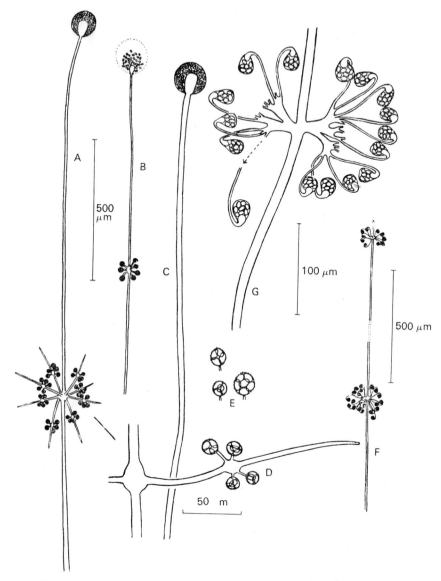

Fig. 3.14. A–E, *Helicostylum fresenii*: (A) sporangiophore with whorl of sporangiole-bearing branches; (B) sporangiophore with whorl of sporangioles, terminal sporangium has become a sporangial drop; (C) sporangiophore lacking sporangioles; (D) sporangiolar branch of A at higher magnification; (E) liberated sporangioles. (F, G) *H. piriforme*: (F) sporangiophore, small straight piece between two whorls of sporangioles omitted; (G) details of sporangiolar whorl, a single sporangiole just liberated is indicated.

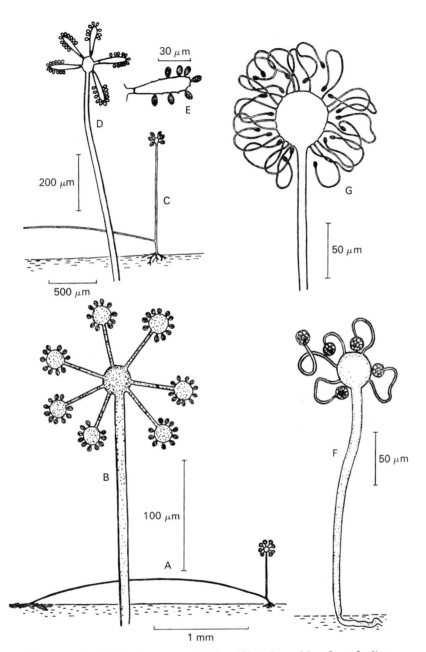

FIG. 3.15. (A, B) *Radiomyces spectabilis*: (A) stolon arising from feeding mycelium on agar producing sporangiophore near region of 'rooting'; (B) details of sporangiophore. (C–E) *R. embreei*: (C) stolon producing conidiophore on agar near its point of 'rooting'; (D) details of conidiophore; (E) attachment of conidia to 'arm' of conidiophore. (F) *Cokeromyces recurvatus*, sporangiophore arising from feeding mycelium on agar. (G) *C. poitrasii*, conidiophore head. (B, D, F, and G in vert. opt. sect).

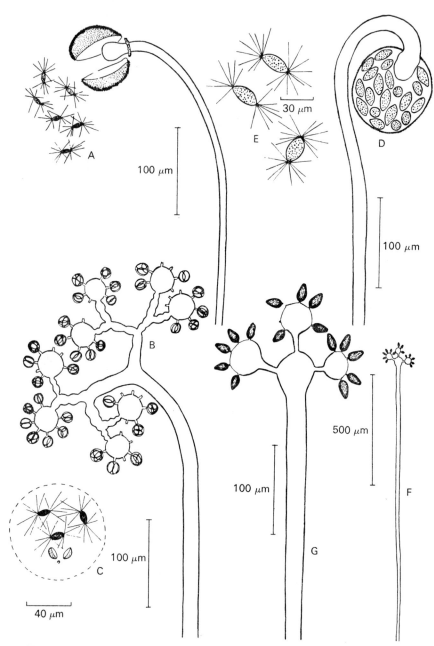

Fig. 3.16. *Choanephora trispora*: (A) sporangiophore in water showing fracture of wall and liberated spores; (B) branched sporangiophore with sporangioles attached to globose heads; (C) a sporangiole that has fallen into a droplet of condensation water on the lid of an inverted Petri dish and has ruptured to liberate its three spores: two halves of old wall of sporangiole and its separated stalk also shown. (D–G) *C. cucurbitarum*: (D) 'nodding' sporangium; (E) liberated spores; (F, G) conidiophore. (B, D, and G in vert. opt. sect.).

In *Piptocephalis* (Fig. 3.17) the long sporangiophore is unbranched below but freely branched in a dichotomous manner in its upper regions. Each branch ends in a specialized 'head-cell' to which the merosporangia are attached. As noted by Benjamin (1959) two types of behaviour may occur. In the first (e.g. in *P. freseniana*) all the spore-rods from one head become involved in a single droplet of fluid, the identity of the individual rods being finally lost. Under the influence of a current of air the head-cells become detached, each carrying its spore-laden drop. In some other species (e.g. *P. virginiana*) the spore rods remain dry. When *P. virginiana* is subjected to a blast of air, some head-cells are blown off carrying their dry spore-rods, but individual rods are also detached and even single spores may become airborne (Ingold and Zoberi 1963).

In *Syncephalis* the simple sporangiophore has a swollen apex from which radiating merosporangia arise. These may be directly attached but more commonly they occur two at a time on a basal cell. Sometimes the contents of this basal cell form what is effectively a spore, so that it seems that a V-shaped merosporangium is involved. This is so for example in *Syncephalis cordata*. In this, and probably in most if not all species of the genus, the head of the sporangiophore becomes involved in slime in which the sporangiospores, liberated by the breakdown of the merosporangial walls, are suspended (Fig. 3.18). In contrast the heads of *Syncephalastrum racemosum* remain dry and the radiating spore-rods are easily blown away (Fig. 3.18E, F). Individual rods tend to remain intact, few spores escaping singly.

In Dimargaritaceae, characterized by two-spored merosporangia, both spore-drop and dry-spore forms occur. In *Dimargaris crystalligena* the tall, unbranched sporangiophore has, as in *Syncephalis*, a globoid head. This bears radiating elongated cells, each the stalk-cell of a branch system. The spores in pairs, each representing a merosporangium, occur in whorls on the shorter, more distal cells of these systems. At maturity all the cells tend to separate from one another except for the large stalk-cells, and the whole head becomes involved in fluid. The result is a stalked spore-drop containing not only spores, which show a tendency to remain in the original pairs, but also the rounded-off cells of the merosporangiferous branch system (Fig. 3.18C, D). Both the sporangiophores of *Syncephalis* and of *D. crystalligena* are further examples of the stalked-spore drop already discussed. However, a dry-spore species of *Dimargaris* has been described (Benjamin 1965).

Species of *Tieghemiomyces*, another genus of Dimargaritaceae, are dry-spore moulds. Two-spored merosporangia are borne in whorls on short, sparingly branched laterals arising at a height of 100–150 μm on an erect hyphae which may be 500–1000 μm long. At maturity, as in other members of the family, the merosporangial walls break down leaving the spores

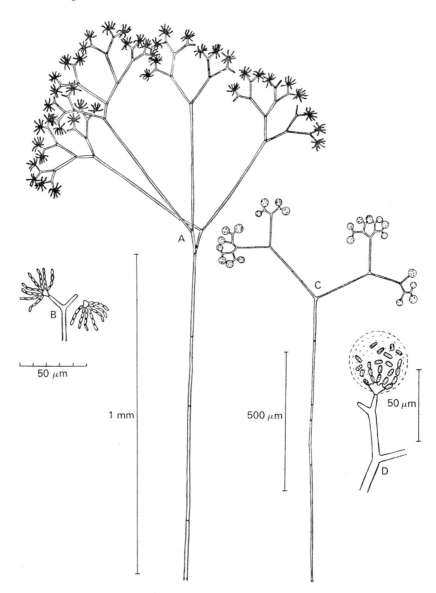

FIG. 3.17. (A, B) *Piptocephalis virginiana* : (A) sporangiophore bearing a tassel
of dry-spore merosporangia at the tip of each ultimate branch; (B) two
head-cells with dry 'spore-rods' (merosporangia) attached; one head-
cell has just separated. (C, D) *P. freseniana* : (C) a rather sparingly branched
sporangiophore with a sporangial drop at the end of each ramulus;
(D) part of the ultimate branching system with a single spore-drop in
optical section.

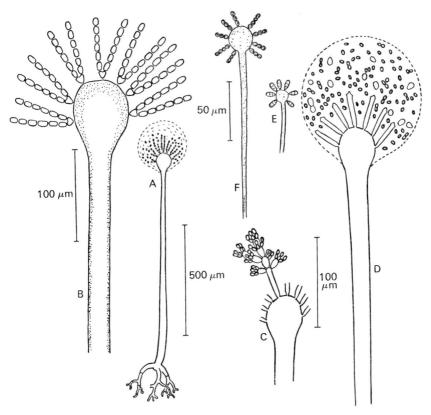

Fig. 3.18. (A, B) *Syncephalis cordata:* (A) sporangiophore with terminal spore-drop; (B) apex of sporangiophore before break-up of the spore-rods; the merosporangial walls have disappeared. (C, D) *Dimargaris crystalligena:* (C) part of apex of sporangiophore showing sporangiferous branches; merosporangial walls have disappeared; (D) later stage with spore-drop formed. (E, F) *Syncephalastrum racemosum:* (E) young sporangiophore with merosporangia in which spores are differentiating; (F) mature sporangiophore with dry spore-rods.

in naked pairs. No slime is produced and the spores are readily blown away, although they tend to remain paired.

Benjamin has drawn attention to a very curious behaviour of the sporulating axis in both species of *Tieghemiomyces*. Immediately below the level of origin of the merosporangiferous branches, the cell of the main axis undergoes a transverse dehiscence. If the projecting part of the main axis is touched with a fine needle or a hair, the whole structure can be lifted off (Fig. 3.19). It appears that the axis is definitely sticky. This remarkable behaviour suggests that a special dispersal mechanism may be involved.

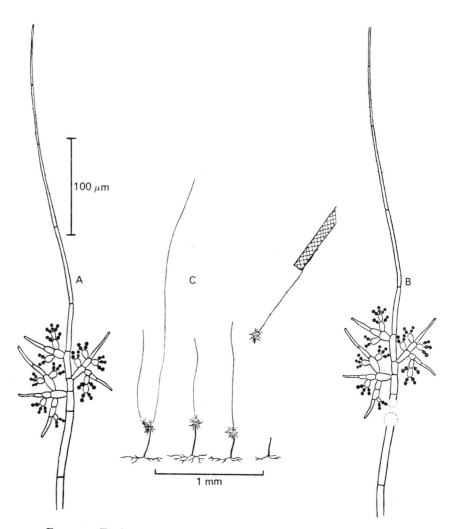

Fig. 3.19. *Tieghemiomyces californicus.* (A) sporangiophore, the two spores
of each merosporangium shown black. (B) the same following rupture of a
cell of the main axis; the free part has been lifted slightly and sap is exuding
from the two halves of the ruptured cell. (C) four sporangiophores arising
from the feeding mycelium, in that on the right the sticky projecting axis
of the sporangiophore has been touched by a human hair (cross-hatched)
and all but the stump has been lifted off.

In some Mucorales the conidial condition occurs. It seems to have arisen, by further reduction, from the sporangiolar and also possibly from the merosporangial condition. This is indicated by certain distinctive genera in which both sporangiolar and conidial species occur. Thus *Cokeromyces recurvatus* has sporangioles while in *C. poitrasii* conidia are produced, or, in other words, the sporangioles have become 1-spored (Fig. 3.15F, G). Other precisely similar pairs are *Radiomyces spectabilis* and *R. embreei* (Fig. 3.15A–C); and *Choanephora trispora* and *C. cucurbitarum* (Fig. 3.16).

Many conidial types are dry-spored and the conidia are easily blown away by wind. Clear-cut examples (Fig. 3.20) are *Chaetocladium* spp., *Cunninghamella* spp., *Mycotypha microspora*, *Radiomyces embreei*, and *Cokermyces poitrasii*. On the other hand, in most Kickxellaceae conidia at maturity become involved in fluid and are not wind-dispersed. In *Kickxella alabastrina* the conidiophore bears a terminal umbel of sporiferous branches each carrying on its upper surface several rows of closely set pseudophialides. Each of these gives rise to a single elongated spore. A drop of fluid is formed that embraces all the spores of an umbel and into this drop they may be liberated. The drop is held in the umbel rather like an egg in an egg-cup and does not encroach on the sporiferous branches which, like the pseudophialides, are unwettable. In *Coemansia* the individual spore-bearing branch has its own droplet (Fig. 3.21).

Spirodactylon aureum, is a remarkable member of Kickxellaceae deserving special mention. According to Benjamin (1959) it is a dry-spore form. The branched conidiophore has a rough spinose surface. At intervals there are spore-bearing regions, each in the form of a fairly tight helix in which spores on pseudophialides face inwards towards the longitudinal axis of the helix. The arrangement seems quite unsuited to normal aerial dispersal (Fig. 3.22). Again, as with *Tieghemiomyces*, it is tempting to suggest that this curious structure is related in some way to the dispersal story of the fungus. Species of both genera grow on mouse dung.

It will be seen from this consideration of a relatively small but probably natural order of fungi that the asexual apparatus shows a considerable range of variation, and it seems that the problem of spore liberation is solved in a number of different ways in the various genera. Further, in some forms closely similar in essential structure, spore liberation follows very different lines. The general moral that may be drawn from a contemplation of Mucorales is that in considering any spore-bearing structure in fungi, it is pertinent to ask the question: how exactly are the spores set free?

A weakness of the account given here of spore liberation in mucoraceous fungi is that it is based largely on study in Petri dish cultures. Conditions in nature may be very different. For example, moulds belonging to this

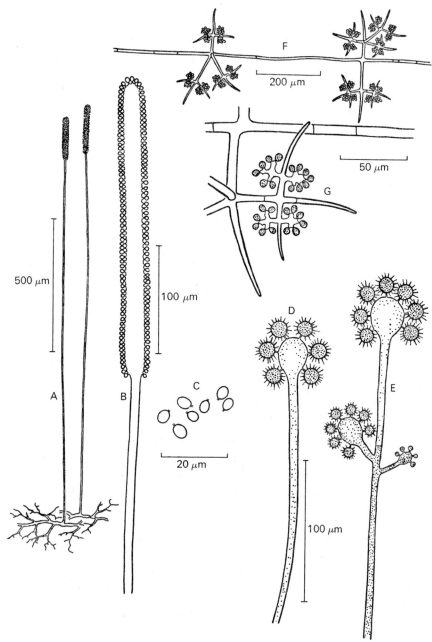

FIG. 3.20. A–C, *Mycotypha microspora :* (A) conidiophore growing up from feeding mycelium; (B) head of conidiophore (vert. opt. sect.); (C) liberated conidia. (D, E) *Cunninghamella elegans*, two conidiophores (vert. opt. sect.) (F, G) *Chaetocladium brefeldii :* (F) part of horizontal conidiophore with lateral sporiferous branches; (G) part of branch more highly magnified.

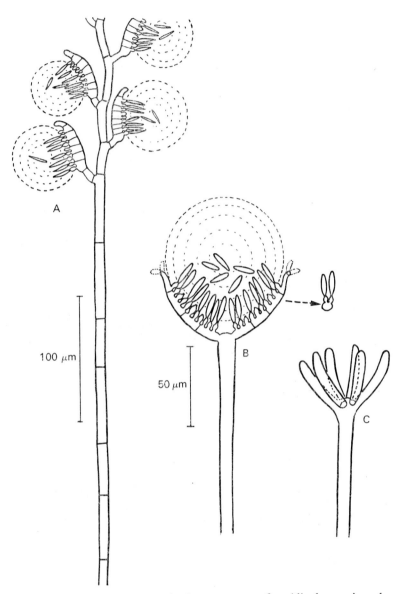

FIG. 3.21. *Coemansia guatemalensis:* upper part of conidiophore axis and lower part of fertile zone (long. sect). (B, C) *Kickxella alabastrina:* (B) mature conidiophore (long. sect. through two sporiferous branches; form of branch tips in dotted outline and appearance in trans. sect. of a sporiferous branch indicated); (C) young conidiophore with umbel of six developing sporiferous branches.

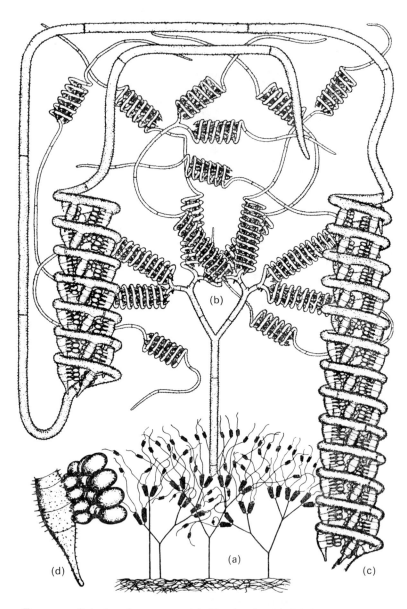

FIG. 3.22. *Spirodactylon aureum*, (*a*) Sketch of general habit of fruiting structures, ×15; (*b*) upper portion of conidiophore showing form of fertile branches, ×150; (*c*) portion at higher magnification, ×435; (*d*) small spore-bearing branch, directed inwards in helix, showing conidia on pseudophialides, ×1575. After R. K. Benjamin (1959).

order are abundant in all kinds of soil, but it is rare under natural conditions to see sporangiophores on exposed earth. Indeed, it is open to doubt if sporangiospores are the chief units of asexual reproduction in soil; perhaps chlamydospores, which may well have a greater capacity to withstand drying, play a relatively important part.

4. DISCHARGE BY ROUNDING-OFF OF TURGID CELLS

RATHER as a drop of water takes on a spherical form under the influence of its surface tension, so also is the tendency to assume a condition of minimum surface found in a living plant cell surrounded by an elastic cell wall. The spherical form may, however, be opposed by local differences in the extensibility of the wall. Nevertheless, as a whole or locally a turgid living cell tends to round off. There are a number of examples, scattered taxonomically throughout the Fungi, in which sudden changes of shape by turgid cells in a state of strain result in violent spore discharge.

This type of spore liberation is found, for example, in Entomophthorales, a small order of fungi mainly parasitic on insects. One of the features of the order is that the terminal conidium is usually shot from its conidiophore. As we have seen in the previous chapter, in *Basidiobolus* and in *Entomophthora muscae* a water-squirting mechanism, involving the bursting of a turgid conidiophore, is responsible for discharge, but in most species of *Entomophthora* and in all species of *Conidiobolus* a rounding-off process is concerned.

Conidiobolus coronatus† seems to be widely distributed as a soil saprophyte, particularly in tropical regions. It also parasitizes termites and aphids. Further, it is known as a human pathogen in Nigeria, causing a subcutaneous phycomycosis of the face (Martinson and Clark 1967). It grows readily on nutrient agar (Martin 1925). The unbranched conidiophore projects from a submerged mycelium and bears a terminal conidium from which it is separated by a cross-wall that bulges into the spore as a hemispherical columella (Fig. 4.1). The limiting wall of the whole structure is two-layered, the outer layer being continuous, enveloping both the conidium and its conidiophore. The actual separating cross-wall is, of course, also double, but one half consists of inner conidiophore wall and the other of inner conidium wall. As a preliminary to discharge, rupture of the outer common wall of conidium and conidiophore occurs around the circular line of their junction. Probably almost at this same moment the incurved region of the inner spore wall bulges outwards and the conidium springs from the conidiophore to a distance of up to 4 cm.

† *C. coronatus* (Cost.) Srin. and Thirum. Syn. *Delacroixia coronata* (Cost.) Sacc. and Syd.; *Entomophthora coronata* (Cest.) Kevorkian; and *Conidiobolus villosus* Martin.

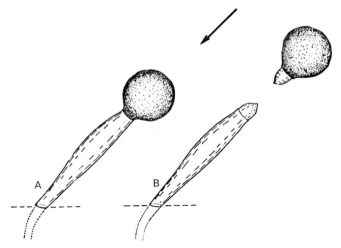

FIG. 4.1. *Conidiobolus coronatus.* (A) conidiophore just before spore discharge, (B) the same at the moment of discharge. Arrow indicates the direction of light. Wall of conidiophore below agar surface dotted. × 500.

A conidium germinates to produce a mycelium only if it falls on a nutrient medium, otherwise, for example on water agar, it germinates at once, usually to produce a very short conidiophore terminated by a conidium only slightly smaller than the parent one. Further, under conditions of unilateral light this secondary conidiophore tends to be formed on the more strongly illuminated side of the original conidium, a photomorphogenic rather than a phototropic response (Plate IV). Sometimes, however, a conidium germinates to produce directly a number (usually more than ten) of conidiophores terminated by very small conidia which seem to be discharged with even greater vigour than the relatively large spores. It is of some interest to note that these small spores (5–10 μm diam) are of a more appropriate size to remain in suspension as part of the air spora (Ingold and Dann 1968).

Weston (1923) has reported a very similar type of spore discharge in the downy mildew *Sclerospora philippinensis* which attacks maize in the Philippines. The mycelium of the fungus is in the intercellular spaces of the leaf, and the branched conidiophores project in tufts through the stomates, being produced only during the night when a film of dew is deposited on the surface of the leaf. Although the lower unbranched parts of the conidiophores are immersed in dew, the upper branched parts bearing the spores project into the air. From each tip a single conidium is formed, the minute surface of contact being flat. Both spores and conidiophore are turgid, so that there is a tendency for each to round-off in this flattened region. Suddenly the adhesive forces are overcome and the base of the spore and the tip of the conidiophore round-off, shooting the spore to a distance of a millimetre or so. Apparently all the spores of a single conidiophore are discharged simultaneously (Fig. 4.2A).

Among downy mildews (Peronosporales) it is only in *Sclerospora* that violent discharge of this nature has been reported, a process involving high humidity. This contrasts strongly with the hygroscopic mechanism found in *Peronospora tabacina* (see p. 130) where discharge depends on sudden drying.

Recently there have been reports of rounding-off processes leading to violent discharge in a number of conidial species that are either imperfect stages of Ascomycetes or fungi classified in the Fungi Imperfecti. Dixon

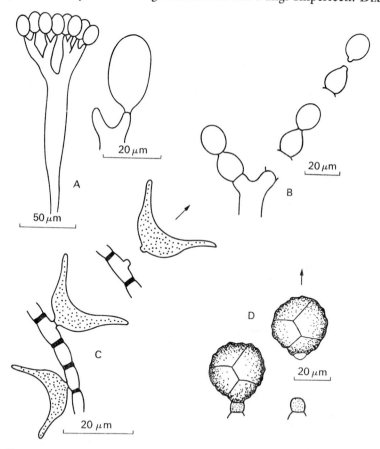

Fig. 4.2. (A) *Sclerospora philippinensis*. A conidiophore, and two ultimate ramuli more highly magnified from one of which the conidium (sporangium) has been discharged. Based on Weston (1923). (B) *Xylosphaera furcata*. End of a conidiophore; *left*, conidium on its supporting cell; *right*, two stages in discharge. Based on Dixon (1965). (C) *Arthrinium cuspidatum*: *left*, part of a conidiophore with two conidia; *right*, conidium discharge involving rounding off. (D) *Epicoccum nigrum*: *left*, attached conidium, *right*, discharge. C and D based on Webster (1966).

(1965) has described the process in *Xylosphaera furcata*, a species habitually associated with the 'sponge' from nests of a common termite (*Macrotermes*) in West Africa. The dichotomous conidiophore of the imperfect stage produces a number of conidia, each rather narrowly attached to an oval supporting cell. When mature, both this and the conidium are discharged to a distance of about half a millimetre by rounding off at the junction between the supporting cell and the rest of the conidiophore. In the air, apparently, the two cells themselves separate, again by rounding-off in the narrow region where they are in contact. Both the conidium and its supporting cell are functional spores (Fig. 4.2B).

Webster (1966) has given convincing evidence that the conidia of *Epicoccum nigrum* (Fig. 4.2D) are discharged by a rounding-off mechanism essentially like that of *Conidiobolus*, and also suggests that spores are shot from the conidiophore in *Arthrinium cuspidatum* in a similar manner (Fig. 4.2C). He remarks, however, that in these two fungi, and in *Nigrospora* (another hyphomycete) in which a water-squirting mechanism is involved, discharge is encouraged by drying. Since turgid cells are concerned, presumably only incipient drying is involved such as may stimulate discharge in certain Ascomycetes.

It is of interest to note, however, that sudden reduction in the relative humidity of the ambient air certainly does not stimulate discharge in *Conidiobolus coronatus*. Indeed, change from high to low humidity causes an immediate reduction in the rate of discharge from an agar culture (Fig. 4.3).

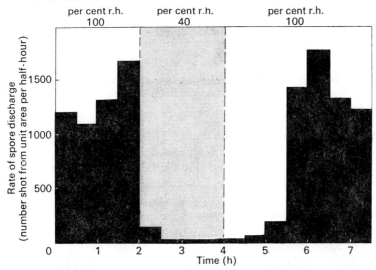

FIG. 4.3. *Conidiobolus coronatus*. Effect on rate of spore discharge (from 1 cm² malt-agar culture) of change in relative humidity from 100 per cent to 40 per cent (stippled period) and back to 100 per cent. Previously unpublished result of an experiment using set-up described by Austin (1968).

Spores are violently discharged from the aecia of rusts, a phenomenon originally described by Zalewski (1883) in a few species of *Uromyces* and *Puccinia*. This has since been confirmed by several workers (see especially Buller (1924)) in many species of *Puccinia* and has also been reported in *Gymnosporangium myricatum* (Dodge 1924). However, I was unable to observe discharge in *G. clavariaeforme*.

Although the force of propulsion is certainly provided by the sudden change in form of turgid cells, the details of the process are not clear. The mature aecium is a minute cup-like structure with the peridium forming a firm wall of thickened cells (Fig. 4.4). At the bottom is a close-set

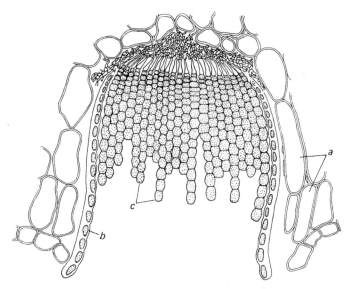

FIG. 4.4. *Puccinia graminis*. Aecium as seen in vertical section of rusted barberry leaf. (*a*) host-tissue; (*b*) peridium; (*c*) chains of aeciospores. × 400.

palisade of basal cells each producing an ever-growing file in which aeciospores and intercalary cells alternate. Each basal cell cuts off a single terminal cell at a time and this then divides to form a larger cell above, which becomes the spore, and a smaller below, which is an intercalary cell. Often the division is by an inclined wall so that the intercalary cell is positioned cornerwise. The intercalary cell remains thin-walled, while the wall of the associated aeciospore thickens. Most workers have described the intercalary cell as an ephemeral structure that soon breaks down. In the cup of the aecium the spores are tightly packed and have polyhedral forms (Fig. 4.4) due to mutual pressure. Under damp conditions the aeciospores absorb water, and with increasing turgor each tends more strongly to become spherical. Spores in the outermost layer of the spore-mass round off suddenly and in doing so spring out either singly or in small groups to a distance of 0·5–1·0 cm.

The release of aeciospores from aecia in the field has been studied in species of *Uromyces* and *Puccinia* (Kramer, Clary, and Haard 1968). As might be expected, discharge occurs mainly at night when the humidity is high or when similar conditions conducive to the turgidity of the spores are produced by rain. It is to be remembered that in these genera the picture is not complicated by a hygroscopic peridium.

In a spore deposit from an active aecium minute spherical germ-plugs often occur both attached to the aeciospores and separate from them (Fig. 4.5), and it has been suggested (Dodge 1924) that these play a part

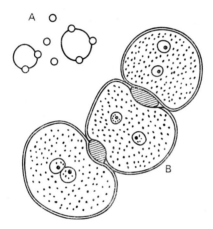

FIG. 4.5. *Gymnosporangium myricatum*. (A) two aeciospores with pore-plugs and three separate plugs (as seen under H.P.). (B) Three adjacent aeciospores and two pore-plugs (more highly magnified). After Dodge (1924).

in the mechanics of discharge. As the aeciospore develops, its wall differentiates into three layers, and at the same time the future germ-pores are organized. In some rusts where a pore is to be formed, the wall is much thicker, and probably different chemically, producing a minute spherical plug that becomes more or less free from the rest of the wall (Fig. 4.5). When the plug is eventually displaced, it leaves a very thin region in the spore membrane through which a germ tube may eventually emerge.

Dodge (1924) suggested an important part in discharge for these germ-plugs. He illustrated his idea by reference to a simple model. A tennis ball compressed over a marble on a table will be thrown further upwards when the confining pressure is suddenly removed than it will be if compressed against the table alone with the same force, and then released. The suggestion is that in the tightly packed aecium the spore is idented by the pore plugs, but in the end the spore suddenly rounds off and is discharged. The pore plug acts as the marble in the model. The major difficulty of this theory is that in some rusts the aeciospores are actively discharged in spite of the absence of pore plugs.

If the spore deposit from a discharging aecium is observed, it will be seen to consist not only of single spores but also to groups of several. Occasionally larger groups of up to 150 spores may be ejected. Buller (1924) refers to these as 'spore bombs'.

Although aeciospores of rusts are normally discharged, urediospores are not, and usually depend on wind for take-off. However, since the uredium of *Coleosporium* spp. is structurally so like an aecium, having its spores in chains, it might be expected that these would be liberated in an active manner; but this expectation is not realized. The urediospores of *Coleosporium* behave like those of other rusts in not being violently discharged.

It has been suggested that ballistospores (including basidiospores of toadstools and the sporidia of rusts) may be discharged by a rounding-off mechanism. It will be seen in Chapter 5 that as a generalization this no longer seems likely, but it must be borne in mind that not all basidiospores are necessarily liberated in the same way. This caution is perhaps necessary, in view of the fact that two quite different mechanisms of violent spore discharge exist (as exemplified by *Peronospora tabacina* and *Sclerospora philippinensis*) within the limits of Peronosporales, a relatively small taxon.

In *Sphaerobolus* the slight but sudden change in form of turgid cells of the palisade belonging to the inner peridium is probably responsible for the spectacular discharge of the glebal mass. It would, therefore, be logical to discuss this fungus in the present context. However, *Sphaerobolus* has been reserved for a later chapter dealing collectively with the remarkable assortment of dispersal mechanisms to be found in Gasteromycetes.

5. THE BALLISTOSPORE

Dr. M. A. Donk gave the name 'ballistospore' (see Derx 1948) to a spore discharged after the manner of the basidiospores in Hymenomycetes, Tremellales, and Uredinales (rusts). The 'secondary conidia' of *Tilletia* and the aerial spores of the mirror-picture yeasts (Sporobolomycetaceae) are also ballistospores. However, in Gasteromycetes and in Ustilaginales (smuts) the spores are not shot from their basidia. Thus not all basidiospores are ballistospores and not all ballistospores are basidiospores.

The characteristic series of visible events associated with discharge of ballistospores is well known in a large range of types due largely to the researches of Buller. His illustration of the process in *Calocera* may be taken as an example (Fig. 5.1). The spore is placed asymmetrically at the

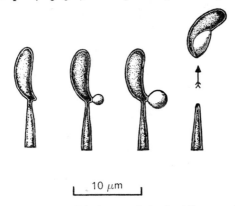

10 μm

Fig. 5.1. *Calocera cornea*. Discharge of the basidiospore (ballistospore) from its sterigma. After Buller (1922).

end of a fine sterigma. Very close to its point of attachment the spore has a minute projection (hilar appendix). Immediately before discharge what appears to be a droplet, but may well be a bubble, suddenly appears at the hilar appendix, grows in a few seconds to a definite size, and then the spore is shot away leaving the sterigma not noticeably altered. The distance of discharge is always short, usually 0·01–0·02 cm and very rarely more than 0·1 cm.

Exact knowledge of the details of the basal part of the mature ballistospore and of its attachment to the sterigma is hard to obtain. The difficulty relates to the small size of the structures involved, and to the fact that when basidia are mounted in water for microscopic examination any really mature spores invariably become detached. Ripe basidia can, of course, be viewed in air, but then only the external features are normally to be seen. One of the morphological problems is whether, immediately prior to discharge, there is a cross-wall separating spore and sterigma. In general, it has proved impossible to decide this matter by light micro-scopy, although Prince (1943) in the rust *Gymnosporangium nidus-avis* considered that there was such a cross-wall and produced a fairly con-vincing photograph as evidence. Wells (1965) has published an electron micrograph of a section through a mature basidiospore and the apex of its sterigma in *Schizophyllum*. The ectoplast membranes of sterigma and spore are separate, indicating that at this stage protoplasmic con-tinuity between the two no longer exists, an electron-transparent region separating the two protoplasts (Fig. 5.2). It appears that the extremely thin outer layer of the sterigma wall is continuous with the outer spore-wall. It is desirable to follow Wells' nomenclature and to distinguish the actual region of attachment as the *hilum* and the characteristic projection near it as the *hilar appendix*, the structure that, above all characterizes the liberated ballistospore. The hilum itself is not usually visible under the light microscope but is readily seen with the electron microscope. It may be a mere scar with a nodulose surface as in *Schizophyllum* (Peglar and Young 1969) or a rather definite projection as in *Sporobolomyces* (Plate V).

Brefeld (1877) first recorded active discharge in Hymenomycetes, but thought that all four spores of a single basidium were liberated at the same time. In the first third of this century Buller made extensive obser-vations, using a wide range of species, and found that successive dis-charge of the spores from a basidium was the rule, although occasionally spores might be set free in pairs. He timed discharge in a number of species. For example, in four basidia of *Stropharia semiglobata* the mean interval between the liberation of the first and fourth spore of the basidium was 302 s, for five basidia of *Coprinus sterquilinus* it was 67 s, and for the same number in *C. plicatilis,* 11 s. It is clear that any adequate explanation of discharge must take account of this successive discharge.

Again Buller repeatedly observed, and this is confirmed by other workers especially Müller (1954), that immediately following discharge the vacated sterigma retains its size and form apparently unaltered, and that no liquid remains behind on its tip. Only as a rare abnormality did Buller see the slow exudation of a small droplet from a sterigma that had just shed its spore.

Initially Buller favoured a rounding-off mechanism at a cross-wall separating the sterigma from the spore, and this theory was later revived by Prince (1943). Such a mechanism is well known in the discharge of the

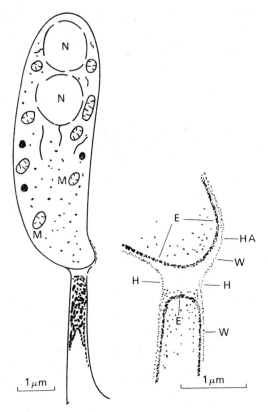

FIG. 5.2. *Schizophyllum commune.* Structure of junction of basidiospore and sterigma. *Left*, at lower power: (N) nuclei; (M) mitochondria. *Right*, part at higher magnification: (E) ectoplast; (H.A.) hilar appendix; (W) wall common to sterigma and base of spore; (H) hilum. Based on electron micrograph by Wells (1965).

conidium in *Conidiobolus* (see p. 92). However, it would not explain the asymmetrical position of the ballistospore on its sterigma, nor would it bring the drop (or bubble), produced at the hilar appendix just before discharge, into the picture. Buller later came to favour a water-squirting mechanism, comparable on a microscopic scale with sporangial discharge in *Pilobolus*; but if the mechanism is of this nature, the sterigma must be self-sealing immediately after discharge, if the turgidity of the basidium is to be retained for the discharge of the subsequent spores. Müller (1954) contributed significantly to knowledge of ballistospore discharge by filming the process in *Sporobolomyces*. In this yeast the 'drop' produced at the hilar appendix grows to full size (3 μm diam, nearly equal to that of the spore) in 2–3 s, before the spore is discharged. However, in one of the fourteen examples, filmed by Müller, the 'drop' (having grown to its normal size) disappeared, within the period of 1/64 s from one 'frame' of the film to the next, leaving the spore behind on its sterigma. He interpreted this as *discharge* of the drop without discharge of the spore. On his view the 'drop' is produced at the *junction* of spore and sterigma and normally is shot (squirted) from the end of the sterigma carrying the spore with it. If, however, what is produced is a bubble, its sudden disappearance could be the result of bursting. Part of Müller's film can, nevertheless, be used as evidence for a drop being formed at the hilar appendix. In his paper (Müller 1954, Fig. 8) two successive 'stills' are reproduced: the first showing the 'drop' grown to its full size; the second the vacated sterigma and, 5–8 μm to the right of it, the discharged spore, just landed on the agar, looking as if it is surrounded by fluid. The effect might, however, be the result of the landed spore being slightly out of focus.

Olive and Stoianovitch (1966) have described a minute slime-mould in which the single aerial spore is apparently jerked off its stalk by the bursting of a gas-filled bubble or blister produced laterally between the outer and inner layers of the spore wall. Figure 5.3 gives a diagrammatic illustration of this organism, *Schizoplasmodium cavostelioides*. The evidence that a gas blister is produced at the side of the spore just before discharge in this slime-mould seems unassailable. Observations on this organism led Olive (1964) to suggest that the mechanism of ballistospore discharge in the true fungi is essentially the same. He considered that in *Sporobolomyces*, for example, the black outlines of the aerial spore and of its sterigma when viewed (growing on agar) under the microscope are due to a gas phase between the inner and outer layers of the wall in these structures. He suggested that a weakening of the outer layer in the region of the hilar appendix leads to this gas being blown out as a spherical blister between the two layers of the wall, and that the ultimate bursting of this leads to the discharge of the spore. It is of interest to note that Weimer (1917), in discussing basidiospore liberation in *Gymnosporangium*, referred to the spore producing a *bubble* at its base prior to discharge, but gave no evidence for regarding it as a bubble rather than as a drop.

The evidence given by Olive (1964) for the presence of a gas phase associated with the fungal ballistospore (e.g. in *Sporobolomyces*) is not completely convincing, resting as it does on the black-edged appearance of the spore, the upper part of the sterigma, and the 'bubble' as viewed in 'dry' mounts under the high-power lens, because almost any hyaline conidium attached to its conidiophore has that appearance if viewed directly surrounded by air. The microscopic appearance of the 'bubble' produced at the hilar appendix does not differ significantly from that of undoubted droplets, for example, those formed on the surface of the sporangiophore of *Pilobolus*. Indeed, it is rather unlikely that a minute droplet surrounded by air would be distinguishable from a corresponding

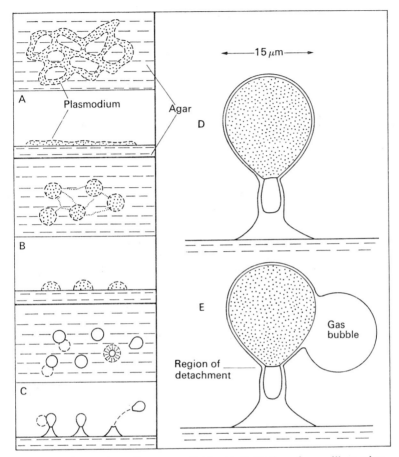

FIG. 5.3. *Schizoplasmodium cavostelioides*. (A, B, C) three frames illustrating development. In each, above is a surface view and below a section of the culture. (A) minute plasmodium of naked protoplasm on agar spread with bacteria; (B) protoplasm has aggregated into hemispherical masses; (C) spores have been produced each on a short stalk; some spores have formed bubbles (outlined by dashed lines); after discharge a crater-like stalk is left. D and E, longitudinal sections of stalked spore on agar. (D) just before bubble formation; (E) at bubble formation. Diagram based on photographs and descriptions by Olive and Stoianovitch (1966).

bubble on the basis of microscopic examination. Nevertheless, the analogy with *Schizoplasmodium* is so suggestive that there is a strong probability in favour of Olive's theory, and it must be remembered that in that slime mould Olive and Stoianovitch clearly demonstrated, in water mounts, the presence of gas between the two layers of wall in the sporulating structure.

It has been argued (Ingold and Dann 1968, Ingold 1969) that under conditions of high external pressure (over 50 atm) a water-squirting mechanism, or one involving the rounding-off of turgid cells, would be able to function, while one involving the blowing of a gas bubble could hardly operate. It was found experimentally that under these conditions ascospore discharge in *Sordaria* and conidium take-off in *Conidiobolus* continued, while the liberation of ballistospores was stopped. In weighing the evidence from these experiments with high pressures, it should be remembered that the actual distance of discharge must be considerably reduced as a result of the increased density of the air. Ballistospores are normally shot to such a short distance (0·1–0·5 mm) that a considerable reduction of this might result in a number failing to be shot far enough for their subsequent escape from the fruit-body. Nevertheless, the immediate, dramatic effect of high pressure on the rate of liberation of ballistospores from sporophores is strong evidence in support of Olive's theory.

If what is produced at the hilar appendix is a bubble limited by an aqueous film, no sign of this would be left after bursting, but if it is a gas blister between outer and inner layers of the wall, a ruptured membrane might be expected in the region of the hilar appendix of the discharged spore. In electron micrographs of whole ballistospores of *Sporobolomyces* no such membrane was detectable (Ingold 1966).

In considering the question of bubble versus drop, it must be recognized that Buller did not merely assume that a drop was carried away with the spore. He attempted to demonstrate its presence on the spore immediately following liberation and before the drop had had time to dry up. His evidence is quite impressive. His principal method was to examine the ripe gill of an agaric covered lightly with a thin cover-slip, leaving here and there an air gap of a fraction of a millimetre between the hymenium and the under surface of the glass. By locating under the high-power lens a basidiospore from which 'drop' exudation was just starting, and then at once focusing upwards on the lower surface of the cover-slip, he was able to see the spore immediately on its impact with the glass a few seconds later. He noted that it was always associated with a little liquid corresponding, apparently, to the exuded drop. Further, Buller reported for a number of species the occasional failure of discharge due to the exuded 'droplets' from the four spores of a basidium running together into a single drop. If bubbles were involved, and especially if they were really blisters, it might be expected that they would retain their identities and not become confluent.

It is of interest that the bubble theory envisages what is essentially a surface-tension mechanism. Such a possibility was originally suggested by Buller and developed by the author (Ingold 1939) who calculated that, on the assumption that a water droplet was formed at the hilar appendix, sufficient surface energy was available to effect discharge. In a bubble of corresponding size twice this energy is provided.

Little is known about the trajectory of discharged ballistospores. Buller's figures (e.g. in *Panaeolus*) suggest that the direction of discharge is parallel with the longitudinal axis of the basidium. If this is so, it is difficult to envisage how a bursting bubble placed laterally at the base of the spore could impart a consistent movement in this direction. Some preliminary observations suggest that the trajectory of discharge is, on the average, at an outward angle to the axis of the basidium (Ingold 1966). The problem of the trajectory worried Savile (1965) and led him to suggest a 'unified theory'. He accepted bubble-bursting as causing the separation of the spore from its sterigma, but thought it must be combined with a 'repulsive' force (possibly electrostatic) acting through the sterigma and propelling the spore along an appropriate course. However, the almost simultaneous action of two violent mechanisms co-operating to secure discharge in the right direction seems highly improbable. Recently Moore (1966) has suggested a way in which the gas pressure in the bubble might be utilized to launch the spore outwards. If, at the moment of break at the hilum, the blister simply collapses driving a jet of gas between the outer and inner walls of the spore in the region of the hilum, not only would there be no ruptured membrane at the hilar appendix, but also a suitable trajectory would be imparted to the spore propelling it in an outward direction away from the sterigma (Fig. 5.4).

Another problem in Olive's theory is the nature of the gas filling the blister. He has suggested that it may be carbon dioxide. In preliminary experiments, however, the author has been unable to observe any decrease in the rate of spore liberation from *Schizophyllum* in the close presence of a

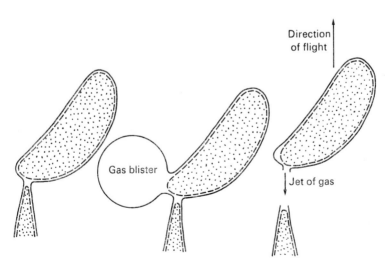

Direction of flight

Gas blister

Jet of gas

FIG. 5.4. *Sporobolomyces*. Diagram of possible behaviour of gas blister during discharge based largely on the ideas of R. Moore (1966).

surface absorbing carbon dioxide. Whatever the nature of the gas it is difficult to see how it manages to accumulate between the two layers of the ballistospore wall.

The problem of the precise mechanism of ballistospore discharge, which has intrigued mycologists for so many years, is not yet solved, but it should be possible in the near future either to confirm or refute Olive's general theory and to clarify the whole position.

Whatever the precise mechanism of ballistospore discharge it seems clear that the cells involved must be turgid. Moreover, release of the spores is particularly sensitive to relative humidity. Using the arrangement shown in Fig. 5.5, involving a colony of *Sporobolomyces* or a fruiting culture

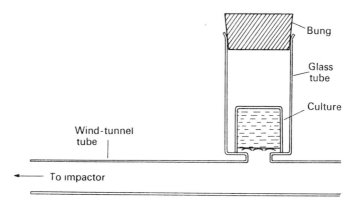

FIG. 5.5. Apparatus used in the study of spore discharge in relation to humidity of air-stream. $\times \frac{1}{2}$.

of *Schizophyllum* in the inverted agar-filled specimen tube, Zoberi (1964) studied the effect of frequent changes from an air-stream of high to one of low humidity. Immediately dry air began to flow below the sporulating surface there was a drastic decline in spore release. There is no suggestion of a temporary stimulation of discharge in response to the onset of dry air that has been observed in some Ascomycetes (Figs. 5.6 and 5.7).

Studies of the air spora also confirm the dependence of ballistospore discharge on high humidity. In the strongly circadian rhythm of the mirror-image yeasts, *Sporobolomyces* and *Tilletiopsis*, the maximum concentration of spores in the air occurs in the very early hours of morning when the air is damp (Gregory and Sreeramulu 1958). A similar picture is found for rust basidiospores as studied, for example, by Carter and Banyer (1964) in *Puccinia malvacearum* and by Snow and Froelich (1968) in *Cronartium fusiforme*. During April and May in the state of Mississippi, U.S.A. the latter workers found very close agreement between the number of hours

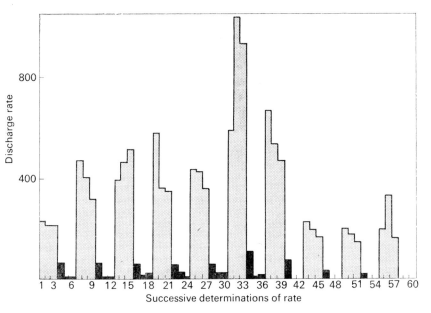

FIG. 5.6. *Sporobolomyces roseus.* Rate of spore discharge and relative humidity. High r.h. (96 per cent) stippled; low r.h. (32 per cent) black; sixty consecutive 5-min observations of rate, three at each humidity.

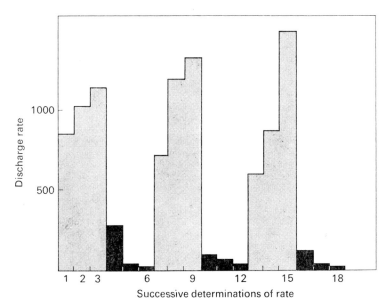

FIG. 5.7. *Schizophyllum commune.* Rate of spore discharge and relative humidity. High r.h. (96 per cent) stippled; low r.h. (56 per cent) black; eighteen consecutive 15-min observations of rate, three at each humidity.

each day when the relative humidity exceeded 97 per cent and the number of hours when basidiospores were present in the air.

Both with the mirror-picture yeasts and with rusts the sporulating structures are directly exposed and rapid response to the local environment might be expected. When, however, the sporulating surfaces are displayed on the closely packed gills of an agaric or lining the long narrow tubes of a polypore, the air bathing the hymenium tends to remain still and saturated, irrespective of fluctuations in the outside air.

6. THE ORGANIZATION OF HYMENOMYCETE SPOROPHORES

HYMENOMYCETES, such as agarices, bracket polypores, and coral fungi, constitute a great assemblage of fungi characterized by the violent discharge of their basidiospores and by having exposed hymenia in the sense that the spores released from their basidia can fall freely and thereby escape from the parent sporophore. The range of fruit-body architecture is limited by certain features of the basidia, namely the short distance (0·01–0·04 cm) of ballistospore discharge, the ability of basidia to function only when the ambient air is close to saturation, and the fact that when mature or nearly so they are ruined by direct wetting.

The toadstool or agaric may be considered first. It is, like other hymenomycete sporophores, essentially a structure concerned with the production and liberation of spores. In most species it is short-lived in contrast to the often perennial vegetative mycelium that is usually hidden within the nutrient substratum. Further, as pointed out by Large (1961), in the total picture of a toadstool there is another phase, the spore cloud. Buller's beautiful drawing (Fig. 6.1) illustrates, for the horse mushroom,

FIG. 6.1. *Agaricus arvensis*. Diagram of the three phases: feeding mycelium in the soil, sporophore, and dispersing spore cloud. After Buller (1909).

the three essential phases: vegetative mycelium, sporophore, and dispersing spore-cloud.

Our understanding of hymenomycete sporophores as functioning structures derives mostly from the morphological and histological studies of living specimens by Buller (1909–31). The structure of an agaric is shown in Fig. 6.2. Covering the gills in most species is a layer of basidia, sometimes interspersed with paraphyses, although in many agarics what have formerly been regarded as paraphyses may simply be young basidia. Further, cells of a very different nature, cystidia, clearly distinct from the other elements of the hymenium and much less numerous, may also occur (Fig. 6.2D).

The basidia are mostly horizontal. Each develops as a club-shaped hymenial cell from the apex of which four fine, curved sterigmata sprout,

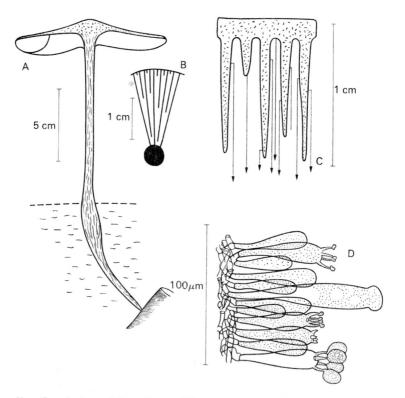

FIG. 6.2. *Oudemansiella radicata.* (A) longitudinal section of sporophore; a pseudorhiza, attached to buried wood, extends upwards into the stipe which bears the pileus. (B) sector of under surface showing gill pattern; the black disk is the stipe seen in transverse section at gill level. (C) tangential vertical section of cap showing vertical gills and the trajectories of some escaping spores. (D) small portion of hymenium with basidia at various stages of development and a single cystidium.

their tips swelling into basidiospores. At first the basidium itself is filled with granular protoplasm but this is driven by the enlarging basal vacuole into the swelling spores, so that in most species the basidium is left with a thin lining layer of enucleate protoplasm (Fig. 6.3). The squeezing of protoplasm into the basidiospores by the piston-like action of the vacuole, described by de Bary (1887) and fully considered by Buller, has been referred to by Corner (1948) as the 'ampoule effect'.

As we have seen in the previous chapter, the basidiospores of Hymeno-mycetes are ballistospores and are violently discharged from the basidium, usually in succession, to a distance of 0·01–0·02 cm. The trajectory of discharge from a horizontal basidium has been considered in detail by Buller (1909) and because of its striking form he gave it a special name: 'sporabola'. This is described by the equation

$$y = \frac{V^2}{g}\left\{-\ln\left(1 - \frac{x}{X}\right) - \frac{x}{X}\right\},$$

where V is the terminal velocity of fall of the spore,

X is the maximum horizontal distance of projection,

g is the acceleration due to gravity,

y is the vertical distance of a point on the trajectory below the point of liberation, and

x is the horizontal distance of this point from the vertical line dropped from the point of liberation.

Essentially, the sporabola consists of an almost horizontal part, completed in a minute fraction of a second, and of a vertical part where the spore falls relatively slowly according to Stokes' law. At first sight the sporobola (Fig. 6.4) seems an improbable trajectory for a discharged object to follow. However, it can easily be imitated by striking a toy balloon horizontally from a table-top. Both with the microscopic spore and with the balloon it is the over-riding importance of air resistance that determines the unusual form of the trajectory.

In considering the first part of the sporabola, it must be borne in mind that there is little information about the exact direction of discharge. It is not known if a basidiospore is normally shot in a direction parallel with the longitudinal axis of the basidium or at an angle to it. Further, nothing is known about the constancy of the direction.

Returning to the general structure of a toadstool, it is clear that the distance between gills must exceed that of ballistospore discharge, otherwise the spores, which are sticky, would lodge on the opposite hymenial surface and escape from the sporophore would be impossible. Also there must be some margin of safety and, in fact, opposing hymenial surfaces tend to be several times further apart than the range of the basidium as a spore gun would seem to require.

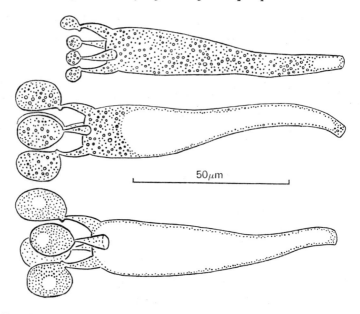

50μm

Fig. 6.3. *Oudemansiella radicata*. Later stages in basidium development. The enlargement of a vacuole in the base of the basidium drives the protoplasm by 'piston' action into the enlarging spores.

Spores shot from the hymenium fall in the still air between the gills and, on emerging below the cap, are well placed for aerial dispersal.

In agarics, with the exception of ink-caps (*Coprinus* spp.), the whole gill ripens simultaneously and at any one time spores are being discharged from most of its surface, although in some species belonging to certain genera (*Panaeolus, Annelaria, Stropharia, Agaricus*) there is a fine-grain mosaic of ripening which does not affect the present argument. Because of this simultaneous ripening, spores may have to fall a significant distance between opposing gills before emerging into the free air below the pileus, and if their escape is to be successful, these must be vertical. This orientation is brought about by growth responses to gravity.

The stipe is negatively geotropic and this ensures that the gills hanging below the cap are roughly vertical. In some toadstools, such as *Agaricus campestris*, this is the only response of the stipe, but in others, particularly in lignicolous and coprophilous species, it is positively phototropic during the earlier stages of development. It had been suggested that as the phototropism dies out it is succeeded by negative geotropism. However, in an analysis of the situation in a stipitate polypore (*Polyporus brumalis*) it has been shown (Plunkett 1961) that throughout development the sporophore is responsive to both light and gravity. The initial

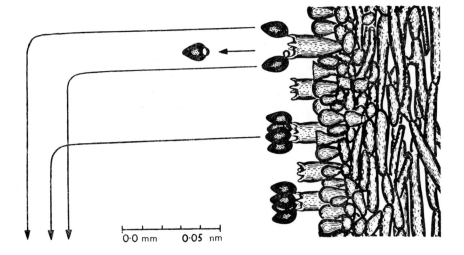

FIG. 6.4. *Panaeolus campanulatus.* Part of the hymenium in vertical section of the gill showing spore discharge. Lines with arrow-heads show trajectories (sporabolas) of spores. After Buller (1922).

phototropism is not lost, but in the later stages the developing pileus shades the perceptive region, the impact of light is thus reduced and response to gravity dominates.

The approximate vertical orientation of gills due to the negative geotropism of the stipe is usually further improved by fine adjustment. The individual gill is positively geotropic. If a pileus is tilted so that the gills are no longer quite vertical, each undergoes growth movements until it is again in the vertical plane. Another feature which in most agarics favours the free fall of spores in that each gill is wedge-shaped in vertical section (Fig. 6.2C).

In a toadstool it is necessary for hymenial surfaces to retain their vertical position from minute to minute. A spore-liberating mechanism of this kind cannot afford to sway in the wind and in this connection the stout stipe has an essential function, for by its solidity it constantly maintains the gills in a correct position for spore release. But the stipe has a further important role in elevating the cap above the ground so that the spores drop into air which is often likely to be in a turbulent condition conducive to dispersal (Fig. 2.4, p. 15).

In the highly specialized genus *Coprinus* (ink-caps), however, the individual gills do not respond to gravity and are parallel-sided, not wedge-shaped, in vertical section. The remarkable organization of the fruit-body in *Coprinus* spp. has been thoroughly studied by Buller (1924).

There is considerable variation in detail from species to species, but
C. atramentarius may be considered as an elegant specific example.

Generally species of *Coprinus* are either coprophilous or lignicolous.
C. atramentarius (Fig. 6.5) is a rather large and common species growing
on buried wood. For the size of the sporophore the amount of pileus
tissue is very small. The gills are remarkably thin (0·1–0·15 mm), very
numerous, and only 0·15–0·2 mm apart. Their hymenia are, however,
prevented from actually touching by turgid cylindrical cystidia that occur
among the hymenial elements of one gill and stretch across to the opposite

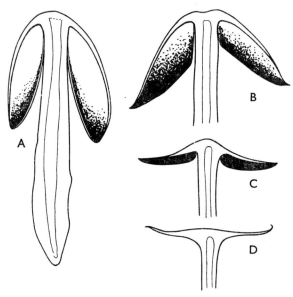

FIG. 6.5. *Coprinus atramentarius.* Sporophore in vertical section. (A) at the
beginning of spore liberation: the free margin of the gill has shed its spores
and is colourless; behind this the gill is black with ripe spores grading off
through pink (dotted region) to white. In B the cap has expanded and a
small part of each gill has been removed by autodigestion. In C most of the
gill tissue has been autodigested, and in D the process is complete. ×½.
After Buller (1924).

one. In most agarics in any small area of the hymenium the basidia are
at all stages of development, but in *Coprinus*, in the high-power field of a
microscope when the gill is seen directly in surface view, all the basidia
are approximately at the same stage projecting from a pavement of
clearly differentiated paraphyses (Fig. 6.6). Further, a wave of development passes from the free margin of each gill outwards and upwards
towards the pileus tissue. Young basidiospores are colourless, but their
walls change as they ripen through pink and dark brown to almost black.

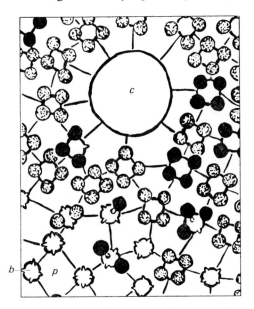

FIG. 6.6. *Coprinus atramentarius.* Small part of the ripe region of the gill in surface view. The base of the picture is near the free edge of the gill and near the bottom are basidia from which all the spores have been discharged. All the spores are black, but those on the shorter basidia have been stippled to distinguish them from those on the longer (shown black). (*b*) basidium which has discharged its spores; (*p*) paraphysis; (*c*) cystidium. × 440. After Buller (1924).

It is the spores that give colour to the gill, so that in a fruit-body from which spores have just started to escape, each gill is blackish towards its free margin and passes through pink to white as the pileus tissue is approached. As the first spores are discharged a very narrow zone near the gill margin becomes white as a result of the loss of its spores, but this spore-depleted area rapidly undergoes autodigestion resulting in its liquefaction. The fluid formed turns black and under humid conditions may flow down the inclined margins of the gills and gather as inky drops around the edge of the cap. However, if conditions are drier the liquid produced by autodigestions evaporates and no inky drops are formed. Most of the spores shot from the basidia escape, but a small proportion drift into the ink and get stuck.

In *C. atramentarius* the stipe is negatively geotropic and this results in the gills being roughly vertical, but the gills themselves are not geotropic. However, their failure to respond to gravity does not militate against the efficiency of spore liberation, since, due to the removal of exhausted hymenium by autodigestion, spores have only a fraction of a millimeter to fall between gill surfaces before emerging into the free air below the cap.

The behaviour of cystidia is especially interesting. Their autodigestion precedes that of the rest of the gill so that they are removed before discharge in their immediate vicinity begins. Thus they offer no obstacle in the path of fall of discharged spores. In this species the basidia are dimorphic, half being relatively long and the rest relatively short. This arrangement allows closer packing in the hymenium than would otherwise be possible. The longer basidia discharge their spores slightly earlier than the shorter ones, and thus ballistospores of the shorter basidia can be discharged without colliding with the spores on the longer basidia (Fig. 6.7). In *C. micaceus* this arrangement is still further developed, the basidia being tetramorphic with the longest basidia discharging first and the shortest last.

Outside the genus *Coprinus* there are a few other agarics with well-defined paraphyses. Buller (1924) describes this condition in *Lepiota cepaestipes* and *Bolbitius flavidus*. In the latter the arrangement of basidia and paraphyses is highly geometric. In a surface view of a gill the paraphyses appear, in a high-power field, like the squares of a chess-board with a basidium inserted at each corner where four squares meet. The basidia are all of the same length, but a quarter of them, spaced as far away from one another as possible, ripen first, followed by a second quarter, and so on until the gill is exhausted, the effective life of the mature fruit-body being only a matter of hours (Fig. 6.8).

Some of the general features of toadstool form merit discussion. Agarics vary considerably in form from minute species of *Marasmius* (e.g. *M. rotula*) to giant species like the horse mushroom (*Agaricus arvensis*).

In solid objects that very in size but not in form three-dimensional features vary as the cube and two-dimensional ones as the square of the linear measurements. Thus on doubling the dimensions of a toadstool the cap volume (and therefore its weight) is increased eight times, but the cross-section of the stipe bearing the load of the cap is merely quadrupled (Fig. 6.9). Adjustments in form with change of size are, therefore, to be expected, otherwise the larger sporophores would have stipes too thin, and the smaller ones unnecessarily thick, for the proper support of the pileus.

An analysis of agarics with central stipes using the data given by Ramsbottom (1923) showed quite clearly the expected tendency for form to vary with size (Ingold 1946). The data given by Ramsbottom are, however, merely a taxonomist's subjective estimates of fruit-body dimensions. A further analysis (Bond 1952) has been made based on the illustrations of Lange (1946), each of which is an accurate drawing of an actual representative specimen. The cap diameter and the width of the stipe near its middle were measured for nearly a thousand species. The average cap diameters of all the figured specimens for each width-class

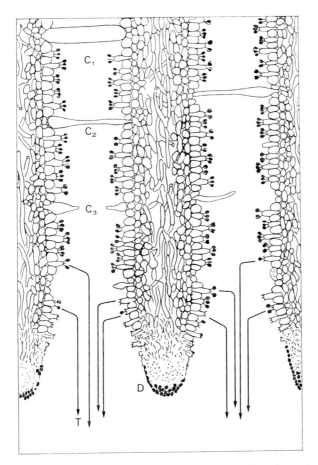

FIG. 6.7. *Coprinus atramentarius.* Tangential vertical section of cap, showing gills in sectional view. C_1, unaltered cystidium; C_2 and C_3, stages in autodigestion of cystidia; (D) autodigesting edge of gill with a few spores trapped in the resulting slime. (T) trajectories of spores being discharged from the basidia. Note the short distance of spore-fall between opposing gill surfaces. × 136. After Buller (1924).

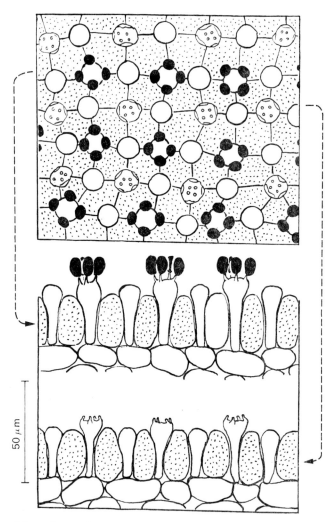

FIG. 6.8. *Bolbitius flavidus. Above*, small portion of gill in surface view showing four ages of basidia: very young, young, mature with spores, and collapsing with sterigmata still visible. First two ages not distinguishable from one another. *Below*, two rows of hymenium at the indicated levels in section: *upper;* young and mature basidia; *lower;* very young, and collapsed basidia. Surface view after *camera lucida* drawing by Buller (1924). Sectional views diagrammatic based on drawings by Buller (1924). Spores shown black (actually rusty yellow). Paraphyses indicated by stippling.

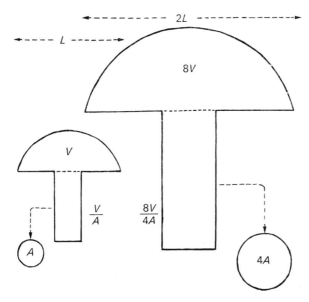

F IG . 6.9. Two toadstool-like objects of identical form but one with double the linear dimensions of the other. In the larger as compared with the smaller the ratio of cap volume to cross section of the stipe is doubled.

of stipe (i.e. 0–1 mm, 1–2 mm, etc.) were determined and plotted against stipe diameter. It is quite clear that the diameter of the pileus does not vary directly as the width of the stipe, but rather that its cube tends to vary as the square of the stipe diameter, as expected on the principle of similitude. This relationship holds not only when all agarics are taken into account, but also when a single large genus such as *Cortinarius* is considered (Fig. 6.10).

Form in toadstools thus shows a tendency to vary in relation to size. Large agarics have relatively stout, and small ones relatively slender, stalks. In general, the stipe is of adequate width not only to support the cap but also to hold it rigidly in a position suitable for spore liberation.

It may be appropriate at this point to consider another feature of fruit-body geometry and, indeed, the rather geometrical form of most sporophores invites considerations of this kind. It has been seen that the gills of an agaric must be a certain minimum distance apart, which may be called the 'safe distance', determined by the range of the basidium as a spore-gun plus a necessary margin of safety probably related to the rigidity of the fruit-body.

If an agaric is examined upside down, it will be seen that the gills converge from the circumference towards the stipe. Only rarely are they all of the same length and stretch the whole distance, although this is the

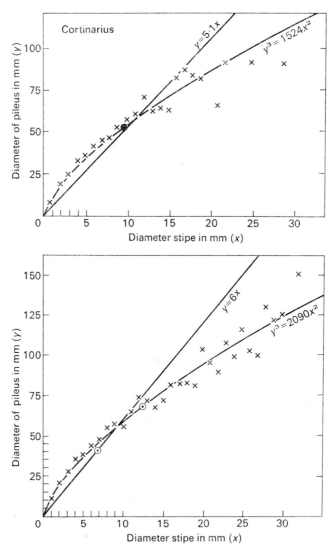

Fig. 6.10. In both graphs the crosses represent diameters of pileus averaged for each stipe diameter (o–1 mm, 1–2 mm, etc). The straight line would be produced if stipe width were proportional to pileus width; the curve is obtained if the cube of the pileus diameter varies as the square of stipe width. Upper graph based on species of *Cortinarius*, lower on all agarics with central stipes. After Bond (1952).

condition in most species of *Russula*. In an agaric it might be considered that near the stipe the gills would be the 'safe distance' apart. If this is so then near the circumference of the cap they would be much further apart then necessary for effective spore escape, so shorter gills could be intercalated between the longer ones. Buller (1909) pointed out that

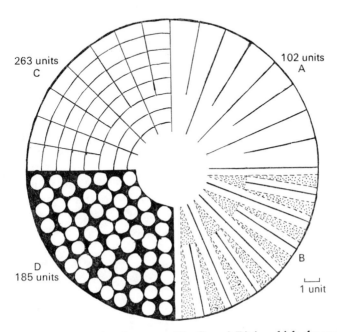

Fig. 6.11. Diagram of various ways (A, C, and D) in which the under surface of a toadstool cap might be partitioned to provide hymenial surfaces. In B the dotted regions show 'wasted' areas when partition is by radial gills. For each type of partitioning the number of units of hymenial area available is indicated.

there tends to be a series of gills of decreasing size expressed by x, x, $2x$, $4x$, etc, where x is the number of full-length gills.

As we have seen, two adjacent gills at their closest approach to the stipe may be considered as being the 'safe distance' apart. Followed outwards towards the periphery of the pileus they become more widely spaced. However, a shorter gill cannot be introduced until the distance apart of the long ones is double the 'safe distance' and only then if its thickness is negligible. From the diagram in Fig. 6.11 it can be seen, therefore, that as a method of partitioning the under surface of the pileus radial gills seem wasteful. It is clear that the available space could be more efficiently used, indeed increased two and a half times, if cross-partitions the 'safe distance' apart were introduced. This arrangement is well on the way to the polypore condition. On one quadrant of the diagram the space has been filled, experimentally not mathematically, by circular pores with the same diameter as the 'safe distance'. Here the extent of the hymenium is almost doubled (Fig. 6.11).

From the foregoing considerations it seems that the polypore arrangement is more efficient in utilizing the area below the cap than a system of gills. An evolutionary pressure from gills towards pores might, therefore, be expected if the production of the maximum number of spores for the minimum expenditure of fungal tissue has selective advantage. It should, however, be observed that in passing from gills to pores the possibility of the fine readjustment of the hymenial surfaces to the vertical position is largely lost. If there really is this supposed advantage, it might be expected that the step from agaric to polypore would have been taken more than once in the course of evolution.

Largely thanks to the work of the modern French mycologists it has become clear that Fries' families Agaricaceae and Polyporaceae cannot be regarded as 'natural' assemblages of fungi. A number of series of what seem to be closely related fungi have been recognized, with agarics at one end, fully developed polypores at the other, and in between intermediate types having anastomosing gills (Fig. 6.12). Two series are especially well documented (Heim 1948):

Mycena, Phaeomycena, Mycenoporella

and

Paxillus, Phylloporus, Boletus.

In any such series the question arises: in what direction should it be read? On the whole it is easier to picture the polypore habit developing from the agaric than the reverse, and if indeed there is greater efficiency in the polypore arrangement, this should argue for selection in that direction. But it must not be assumed that the polypore has always arisen from agaric ancestors. Indeed, it is highly probable that the shallow-pored forms have developed directly from the thelephoroid types.

The essential organization of fleshy toadstools without gills resembles that of agarics. Thus in *Hydnum repandum* the individual downward projecting hymenial teeth are positively geotropic and Buller claims that this is also true of the pores of *Boletus*. He points out that, like the gills of agarics, the tubes of *Boletus* form under the cap as the result of a morphogenetic stimulus and not under the stimulus of gravity as in a species of *Polyporus*. Only later, when the cap has expanded, do the tubes react to gravity to make the small adjustment to bring the hymenial surfaces into the vertical position. However, it must be borne in mind that room to manoeuvre of gills and teeth in this final adjustment is much greater than with *Boletus* pores, for these are relatively long tubes united into a coherent system.

Brief mention may be made of the coral fungi classified in Clavariaceae. In these fungi the hymenium covers the upper parts of either a club-shaped unbranched sporophore or a branched sporophore with vertical

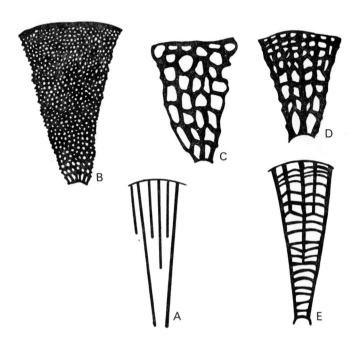

FIG. 6.12. Sectors of under surface of cap in a number of fungi closely related to *Mycena* (the condition in that genus being shown diagramatically in (A)). (B) *Mycenoporella clypeata*; (C) *Poromycena manipularis*; (D) *Phlebomycena madecassensis*; (E) *Phaeomycena aureophylla*. (B, C, D, and E) after Heim (1948).

branches. Corner (1950), whose monograph of these fungi is one of the classics of mycology, points out that species of very different lineages are probably grouped together in this Friesian family. In this connection it is worth remarking that the common fungus *Calocera viscosa*, so abundant on dead coniferous stumps, has all the appearance and organization of a *Clavaria*, but is classified not only outside Clavariaceae but outside Hymenomycetes on the basis of its Y-shaped basidia and its gelatinous texture. The clavarioid sporophore may probably be regarded with confidence as a rather primitive type if only for its failure to provide shelter from rain for the hymenial surfaces, for a hymenium of basidia is spoilt, at least temporarily, if wetted. The form of a toadstool or a bracket fungus protects hymenia efficiently from rain, but in *Clavaria* there is no provision against this hazard.

The general features of the organization of bracket polypores are essentially similar to those of fleshy hymenomycetes, but there are interesting differences. *Polyporus betulinus* may be taken as an example (Fig. 6.13). In this very common parasite of moribund birch which, having killed its host produces its fruit-bodies on the trunk and limbs of the dead tree, gravity has a profound formative influence on the sporophore.

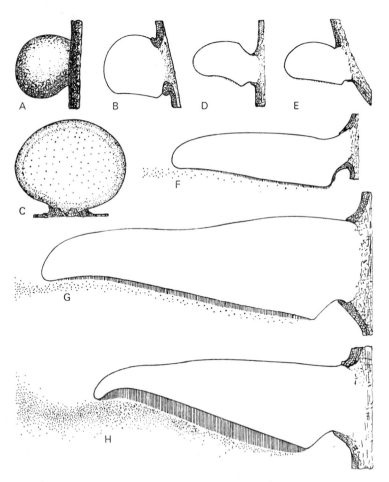

FIG. 6.13. *Polyporus betulinus.* (A–H,) sporophores at increasing stages of maturity. All, except (A) and (C) in vertical section, (C) being a surface view from below. (A) is the almost spherical primordium before any flattening of the under surface has occurred. At (F) tube-formation has just begun but already spores are being liberated. After Buller (unpublished).

The young fruit-body first appears in late summer as a small, spherical knob 2–3 cm across. If this is on the main trunk, it then grows out more or less horizontally (diageotropism) to form a semicircular structure with a firm brownish 'cuticle' above and a smooth softer and pure white surface below. From this the hymenial pores develop. At first they are very shallow, but throughout this part of the life of the sporophore (September to April) they grow by means of an active zone around the mouth of each pore, so that gradually they become longer. From their first inception the hymenial pores liberate spores. The downward direction of pore growth is conditioned by gravity so that the tubes produced are precisely vertical. This gradual geotropic growth contrasts sharply with what happens in a fleshy agaric or in a bolet where gravity is concerned not with the general morphogenetic positioning and development of the hymenial surfaces, but only with their final orientation. In bracket polypores geotropic growth achieves the desirable result of vertical hymenial surfaces, but if these, once formed, are slightly displaced from the vertical, there is no mechanism of readjustment.

Consideration has been given to a sporophore of *P. betulinus* developing on a vertical trunk. If the original spherical primordium is on the under-side of a horizontal branch, the fruit-body grows in a roughly circular form with a central attachment. Sporophores do not normally arise on the upper side of a branch, but if a dead tree bearing primordia is felled, those on the recumbent trunk may continue to develop. A primordium exposed on the upper surface grows out, normally on one side only, and from the underside of the extension the vertical tubes arise (Fig. 6.14).

The leathery, corky, or even woody texture of bracket polypores differs strikingly from that of fleshy agarics and they liberate their spores over a longer period. The life of a toadstool is normally reckoned in days; that of a bracket polypore in months. Again, most of these fungi are xerophytes losing water in dry weather and ceasing to discharge spores, but capable of rapid absorption of water when wetted by rain followed shortly afterwards by renewed spore release. The great majority of these fungi, however, have fruit-bodies that last for only a single season. In striking contrast, however, are a few large, woody forms with perennial sporophores, the most outstanding examples being *Fomes fomentarius* and *Ganoderma applanatum*.

G. applanatum is frequently to be seen on the trunks of broad-leaved trees especially beech (Fig. 6.15). As in *Polyporus betulinus*, gravity has a determining formative effect on fruit-body development. Because of its rigid woody structure and because of its broad and firm attachment to the tree-trunk, the hymenial tubes (0·01–0·02 cm in diameter) formed by geotropic growth are very accurately vertical and unlikely to suffer subsequent displacement. The tubes increase in length during the growing season (May to September) and then rest until the following year when

growth is resumed. Thus annual layers of tubes are formed. If a sporophore is cut vertically with a saw, it is not easy to distinguish boundaries
between adjacent layers. If, however, a deep incision is made in the upper
sterile tissue, and the fruit-body is then broken in two by bending, the
vertical surface exposed has a stepped appearance, each step being a
layer of tubes. Some structural peculiarity associated with the cessation

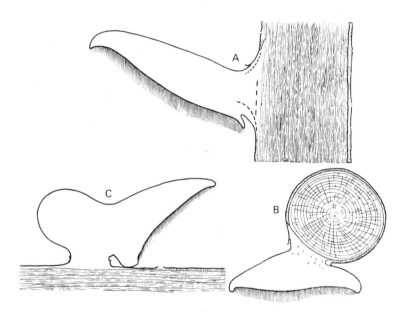

FIG. 6.14. *Polyporus betulinus*. (A) L.S. through trunk of small birch tree,
showing sporophore; (B) T.S. horizontal branch of birch, showing sporophore on under side; (C) sporophore, on upper side of felled birch trunk,
developed from primordium which arose when the trunk was erect. × ⅖.

of growth each year is, no doubt, responsible for this tendency of one
annual stratum to break away from that above and below, in spite of the
fact that they are not separated by layers of sterile tissue. The hymenium
remains active in the tubes for several years and in consequence a spore
from a hymenium, which originated three or four years earlier, may, on
discharge into a tube only 0·01 cm wide, have to fall a distance of 5–10 cm
before emerging into the free air below the sporophore. It might be
expected that many such spores would become stranded on the hymenial
surface as they fall, but direct evidence suggests that this occurs to a
very slight extent. The distance of horizontal discharge in *G. applanatum*
appears to be 0·005–0·01 cm, but this range is based on rather few observations. It is clear, however, that there is almost no margin of safety, a

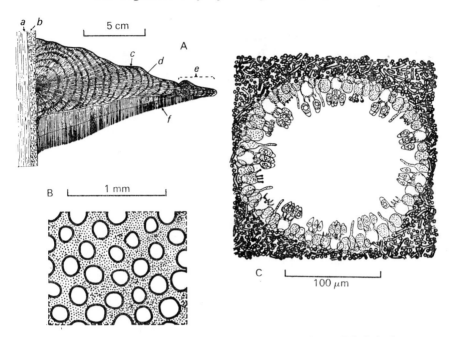

Fig. 6.15. *Ganoderma applanatum.* (A) vertical section of a small fruit-body growing on an ash tree: (*a*) wood; (*b*) bark; (*c*) upper crust of fungus; (*d*) zoned fibrous pileus tissue formed in the first year; (*e*) additional pileus tissue produced in the second year; (*f*) hymenial tubes (the dotted line indicating the boundary between the first and second year of growth). (B) horizontal section of fruit-body at the level of the hymenial tubes; the thick black line around each pore being the hymenium (C), details of a single hymenial pore.

state of affairs to be contrasted with that in most agarics where the space between opposing gills is usually several times that of the discharge distance.

The accuracy of tube adjustment required, combined with the considerable distance spores may have to fall before emerging into the free air, has lead to the idea that there might be some additional mechanism helping to prevent the stranding of spores on the sides of the tube. It was suggested (Ingold 1957) that electrostatic charges on the discharged spores might keep them off the hymenium. Much earlier, Buller (1909) demonstrated that basidiospores on liberation from a sporophore are habitually charged, and this was confirmed by Gregory (1957) who showed that spores emerging below a bracket of *Ganoderma* mostly carry a positive charge. Further work (Swinbank, Taggart, and Hutchinson 1964) using *Merulius lacrymans*, and involving the determination of the actual magnitude of the charges, has, however, indicated that these are minute and are unlikely to be in any way concerned with the escape of spores from hymenomycete sporophores. Again, if any repelling action is to occur to keep the spores falling down the middle of the tube, a charge of the same sign is needed on the hymenium, and there is no evidence that this exists. Moreover, there is now some good evidence that fall under gravity is the only factor that need be invoked.

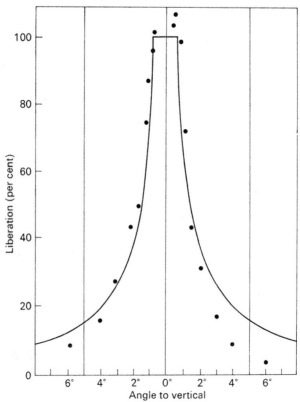

FIG. 6.16. *Polyporus betulinus*. Solid line shows theoretical curve for liberation of spores, as a percentage of that from the vertical position, from a hymenial tube 6 mm long and 0.08 mm in radius, assuming that spores are being discharged at the same rate over the whole hymenium and are shot to the centre of the tube before falling. Dots represent averages, based on actual measurements of liberation from four hymenial tubes roughly circular in section and of the mean dimensions given above. Based on Taggart *et al.* (1964).

The effect of tilting on the liberation of spores has been studied by Taggart, Hutchinson, and Swinbank (1964) in *Polyporus betulinus*. Their elegant set-up allowed spore escape to be measured from individual hymenial tubes at increasing angles of deflection from the vertical.

They found that the reduction in spore escape on tilting was fully explic-
able on purely geometrical considerations, assuming that the spores were
shot into the middle of the hymenial tube and then fell under gravity.
The tubes were 0·016 cm in diameter and 0·6 cm long. A 5° departure
from the vertical reduced spore escape to about 15 per cent of the value
when the hymenial tubes were exactly vertical (Fig. 6.16). With the
much longer and somewhat narrower tubes of *Ganoderma*, tilting would
have an even more disastrous effect.

7. DISCHARGE CONNECTED WITH DRYING

VIOLENT spore discharge is not merely a feature of terrestrial fungi; it also occurs in cryptogamic green plants, especially in polypod ferns and in a number of bryophytes, particularly leafy liverworts, *Sphagnum*, and a few other mosses (Ingold 1965). In these flowerless land plants, however, the mechanisms of discharge depend on drying, whereas in most fungi spore discharge usually involves the activity of turgid living cells, so that damp conditions are needed for sustained spore liberation. However, there are a few instances of active release in fungi where the mechanisms are operated by drying. Two types can be recognized: first, those in which hygroscopic movements are involved; secondly, those in which the sudden rupture of stretched water releases the mechanism of discharge.

Movements, often involving spectacular twisting connected with the collapse of sap-filled hyphae on sudden drying, may result in the violent discharge of spores associated with these hyphae. The best-known example is the liberation of spores from the branched sporangiophores of *Peronospora tabacina* (Pinckard 1942) and the same mechanism probably operates in some other downy mildews, although precise observations are largely lacking.

In *P. tabacina* (Fig. 7.1) the non-septate sporangiophore rapidly loses water on exposure to dry air and becomes twisted around its longitudinal axis. This twirling movement, suddenly executed, has the effect of throwing off the finely attached spores. Violent movements also occur when the partially dried sporangiophore is suddenly subjected again to damp air. However, the evidence suggests that in nature liberation of spores occurs only when the change is from dampness to relatively dry conditions. Normally in downy mildews a fresh crop of sporangiophores matures overnight on the diseased leaves and the conidia (sporangia) are liberated in the morning with the sudden fall in humidity soon after sunrise.

This mode of discharge was recognized many years ago by de Bary (1887) who thought that it occurred in a number of conidial fungi. More recently Jarvis (1962) has considered how far it operates in the common grey mould, *Botrytis cinerea*, attacking ripe fruit in a raspberry plantation. He concludes that, in general, hygroscopic movements release spores from organic contact with the conidiophores, but with insufficient

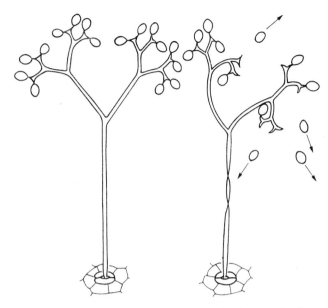

FIG. 7.1. *Peronospora tabacina*. *Left*, turgid conidiophore under damp conditions. *Right*, the same a few seconds later after exposure to dry air (highly magnified). After Pinckard.

vigour to cause actual discharge. Such released spores, in contrast to those still attached, are in a condition readily to become airborne as a result of wind action or agitation caused by rain.

Another hygroscopic mechanism leading to vigorous scattering of spores occurs in some slime moulds (Mycetozoa), especially in the genus *Trichia*. In *T. persimilis*, for example, the thin papery wall of the mature sporangium eventually ruptures in an irregular manner to expose a mass of spores and elaters (Fig. 7.2). These show a striking resemblance to the elaters of liverworts, although their origin and development are totally different. With change from damp to dry conditions, they twist vigorously and stresses are set up between neighbouring elaters which may be suddenly relieved by their ends springing free thereby flinging adhering spores into the air. The elaters of *Trichia* behave like those of the liverwort *Marchantia*, becoming strongly distorted in the dry state but straightening out in damp air. There is no ultimate separation of a gas phase as in the elaters of leafy liverworts (see Ingold 1964).

Although hygroscopic discharge is not involved, it may be appropriate here to refer to the escape of aeciospores from the aecia of *Gymnosporangium juniperi-virginianae*, which is controlled by a hygroscopic peridium (Fig. 7.3). Pady and his co-workers (1968) have demonstrated a diurnal periodicity in spore liberation from apple leaves attacked by this

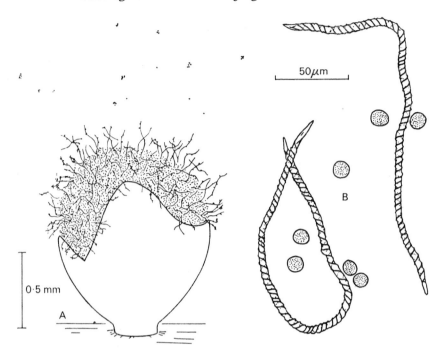

FIG. 7.2. *Trichia persimilis.* (A) sporangium standing on rotton wood. Upper part of sporangial wall eroded exposing spores and elaters. Single spores and groups are being thrown into the air due to movements of drying elaters. (B) elaters and spores.

rust, the daily peak being clearly related to a reduction in the relative humidity of the air (Fig. 7.4). In the aecium the emergent part of the peridium is dissected longitudinally into a large number of filaments, each consisting of a longish row of peridial cells, every one of which has its inner wall thickened while the outer remains thin. The result is that drying leads, as in the fern annulus, to the filaments bending backwards. Thus when the air is damp ($>$ 90 per cent r.h.) the peridial filaments cover the spore-mass in the aecium, but when it is drier they swing outwards exposing the aeciospores which are thus free to escape only when the relative humidity is reduced below 90 per cent r.h. (Pady, Kramer, and Clary 1969*b*).

It is interesting to note that the distribution of thickening in the peridial cells of *G. juniperi-virginianae* is just the opposite of that in aecia with non-hygroscopic peridia such as occur in species of *Puccinia* and *Uromyces*. In these rusts the daily peak of discharge is associated with periods of high humidity when the aeciospores are turgid (see p. 97).

To understand discharge mechanisms released by water rupture, it is first necessary briefly to consider water in the stretched tensile condition.

For this purpose the most instructive demonstration of the behaviour of water under increasing tension comes from a consideration of what happens in Berthelot's experiment (Dixon 1914) which is not too difficult to perform.

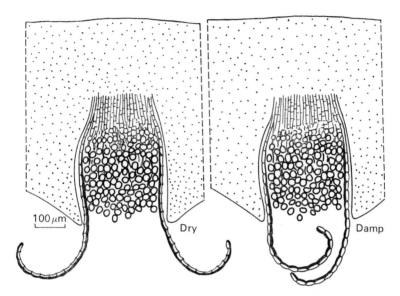

Fig. 7.3. *Gymnosporangium juniperi-virginianae.* Part of T.S. apple leaf showing vertical section through aecium under dry and damp conditions. Host tissue shown by stippling.

A thick-walled glass tube, some 25 cm long and with an internal diameter of about 0·1 cm, is thoroughly washed to remove any grease and all particles of dust. It is then sealed at one end, and drawn out and bent over at the other to form a fine hook the open end of which can dip into a small reservoir of water. Then, by alternate heating and cooling, the tube can be completely filled with water at, say, 30°C. The tube is then removed from contact with its reservoir, and allowed to cool to 20°C, and as the water contracts a very small volume of air is drawn into the tube. Its open extreme tip is then sealed in a flame. If the tube is now heated to 30°C or rather more, the water again expands completely filling it, the air being forced into solution. If a bubble remains, even one only just visible with a strong lens, this simply enlarges if the tube is again cooled. However, provided the air has completely disappeared, on cooling the water continues to fill the whole tube even if the temperature falls below 10°C. The water is then in a state of tension; it is stretched, and this is possible because of the cohesion of the water molecules and of their adhesion to the glass walls of the tube. The tube itself, in spite of its

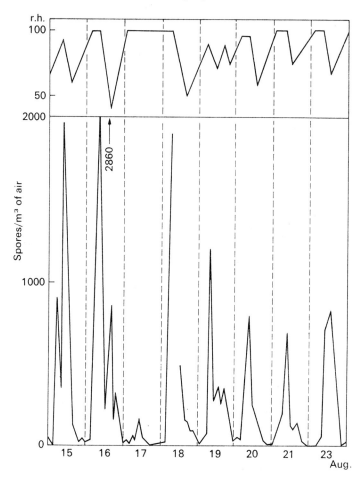

Fɪɢ. 7.4. *Gymnosporangium juniperi-virginianae. Below:* concentration of
aeciospores in air (sucked from the immediate neighbourhood of apple
leaves bearing aecia in August) plotted against time. *Above:* graph of
relative humidity (r.h.) during the same period. Interrupted vertical lines
indicate midnight positions. After Pady *et al.* (1968).

thick wall, is no doubt also distorted to some extent. As cooling proceeds
below 10°C, suddenly, and with an audible tinkle, the water ruptures
and minute gas bubbles appear which soon run together into one, the
water contracting to its normal volume at the particular temperature
involved (Fig. 7.5).

In a thick-walled cell, which cannot easily be distorted, evaporation
may lead to the aqueous contents passing into a state of tension, and
finally water rupture may occur, as in a Berthelot tube, with the separation
of a gas phase, probably water vapour. This process can readily be

observed if the thick-walled spores of *Podospora* or *Sordaria* are mounted in a thin film of water and watched under the microscope as drying proceeds (Ingold 1956). The spore wall is too rigid to allow significant distortion and very soon after the external water has evaporated, a gas bubble suddenly appears within the spore. If spores in this state are again flooded with water, the gas bubbles disappear in a few seconds (Plate VI). If the wall of the cell in question has thin regions, the effect of

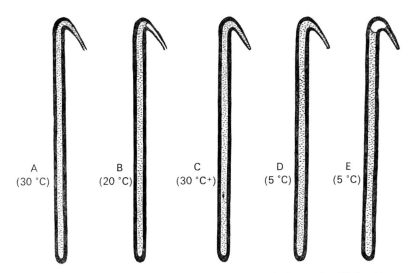

A   B   C   D   E
(30 °C)   (20 °C)   (30 °C+)   (5 °C)   (5 °C)

FIG. 7.5. Berthelot's experiment. (A) thick-walled glass tube filled with water at 30°C, tip open; (B) cooled to 20°C and then extreme tip sealed in flame; (C) heated to over 30°C, water again completely fills tube; (D) now cooled to say 5°C, water still filling tube; (E) same temperature but water-rupture has occurred and water has contracted to its normal volume at 5°C and a gas phase has appeared.

drying is a distortion of the cell, by a drawing inwards of the thin parts, as the water within passes into the stretched tensile condition. When the water breaks and a gas phase appears, the cell returns in a flash to its original volume with the consequent immediate enlargement of the gas bubble. This is the principle involved in the violent movements of elaters in leafy liverworts and of the back-flick of the distorted fern annulus which lead to spore discharge in these cryptogams, and Meredith has reported spore liberation of essentially the same nature in no less than ten genera of dematiaceous Hyphomycetes (Meredith 1961; 1962a, b; 1965). Two typical examples may be considered.

The series of events associated with the discharge of the conidium in *Deightoniella torulosa* is illustrated in Fig. 7.6A. The terminal cell of the conidiophore is thick-walled, but the thickening is not equally deposited,

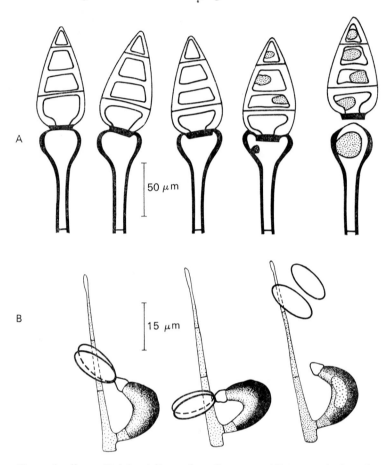

FIG. 7.6. *Above: Deightoniella torulosa.* Septate conidium attached to the apical cell of conidiophore; stages in drying leading to spore discharge; gas-phase stippled. Only a minute fraction of a second separates the last two stages. *Below: Zygosporium oscheoides.* Changes on drying associated with spore discharge: The gas phase which finally appears in the 'falx' is shown white. Both from Meredith (1961, 1962).

being much less in the apical region, and consequently a change of form occurs on drying, the reduction in volume as a result of evaporation being achieved by the thin-walled apex being drawn inwards. At this stage the watery contents are in a state of growing tension. Eventually the water breaks, a gas phase makes its appearance, and in the same instant the apical cell of the conidiophore returns to its original form discharging the conidium in so doing.

Another example is found in *Zygosporium oscheoides* (Fig. 7.6B). In this the short conidiophore has a specialized, curved cell, referred to as a 'falx'. This has a dark, almost opaque, wall strongly thickened on the convex but relatively thin on the concave side. Apically it bears two small thin-walled cells each attenuated into a fine sterigma bearing a small oval spore. When exposed to dry air, evaporation leads to increased curvature of the 'falx'. Finally, water-rupture occurs and then the falx immediately straightens out and returns to its former shape jerking the spores off their sterigmata in the process.

In addition Meredith has observed discharge of essentially the same nature in eight other genera: *Alternaria, Cordana, Corynespora, Curvularia, Drechslera, Helminthosporium, Memnoniella,* and *Zygophiala*.

8. GASTEROMYCETES, OR NATURE TRIES AGAIN

CONSIDERING the differences between Hymenomycetes and Gasteromycetes, the essential feature is that in Hymenomycetes the basidia are displayed on hymenial surfaces which are so disposed that when they discharge their spores these fall freely from the sporophore. In Gasteromycetes, however, when the basidia are mature the hymenial surfaces, which are often poorly defined, are not exposed and the individual spores are not violently discharged. Apparently correlated with this, the basidiospore is not poised asymmetrically on a curved tapering sterigma, but is either arranged symmetrically on a sterigma which rarely tapers or is sessile (Fig. 8.1). At maturity in most Gasteromycetes the basidia break down and the spores lie freely within the ripe sporophore.

The classification used in this chapter is that given by Ainsworth (1961). In this Gasteromycetes are subdivided into a number of orders: Hymenogastrales, Phallales, Lycoperdales, Sclerodermatales, and Nidulariales. Hymenogastrales include several families especially Secotiaceae, and Hymenogastraceae which is a family of specialized hypogeal fungi.

Although for convenience most mycologists retain the nineteenth century concept of Gasteromycetes, it has become increasingly evident that this taxon cannot reasonably be regarded as a natural one. Largely due to the work of Heim, Malençon, and Romagnesi in France, and of the American mycologists Singer and Smith, a number of suggestive series have been identified which seem to link gasteromycete genera with others amongst Hymenomycetes. In seeking these links most reliance has been placed on characters that do not appear to have adaptive significance such as shape, colour and ornamentation of spores, and the details of hyphal construction of the fruit-body.

Certain genera of Gasteromycetes are unmistakably linked with toadstools. A striking example is the genus *Secotium* with some twenty-five species. *S. agaricoides* in its early development agrees exactly with *Agaricus*. However, the gills anastomose considerably and the fruit-body does not open to expose hymenial surfaces from which spores are shot. Instead, at maturity the spores are set free passively within the sporophore and most of the gill tissue breaks down. Eventually the fruit-body cracks in rather a rough and irregular manner and the dry spores sift out (Buller 1922).

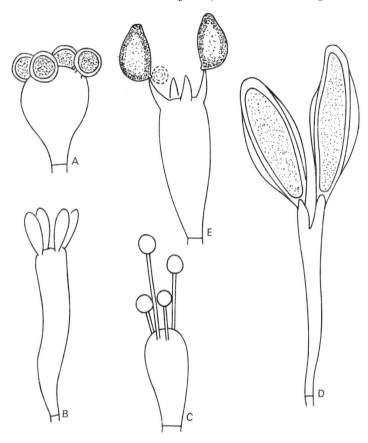

FIG. 8.1. Basidia of Gasteromycetes: (A) *Scleroderma aurantium*, (B) *Phallus impudicus*, (C) *Lycoperdon perlatum*, and (D) *Hymenogaster citrinus*. Also (E) the basidium of the hymenomycete *Panaeolus campanulatus* showing two spores still attached one of which is on the point of discharge. All highly magnified.

On the basis of the peculiar 'crystallographic' spores combined with their pink colour Romagnesi (1933) sees a clear connection between the agaric genus *Rhodophyllus* and the hypogeous gasteromycete *Richoniella*.

Having regard to details of histology, especially the presence of lactiferous hyphae and nests of sphaerocysts combined with amyloid spores, Heim and Singer and Smith (1960) see an evolutionary line between toadstools of the *Lactarius-Russula* type and such secotiaceous genera as *Elasmomyces* and *Archangiella*.

Other plausible connections are between the ink-cap *Coprinus* and the black-spored gasteromycete *Gyrophragmium*, between *Galera* in agarics and *Podaxis* in Gasteromycetes, both with the same type of brown spore having a large terminal germ pore, and between *Boletus* and *Rhizopogon*, both attacked by the same rather specialized mould, *Sepedonium chrysospermum*.

There are at least half a dozen distinct and fairly convincing connecting series between Gasteromycetes and fleshy Hymenomycetes. However, it is worth noting that all these connections are with the secotiaceous types within Hymenogastrales, only one of the five orders of Gasteromycetes. Good evidence of links between the remaining four orders and Hymenomycetes is non-existent, nor are there any obvious lines joining Hymenogastrales with other Gasteromycetes. From the point of view of a *natural* classification it might be best to shunt the order Hymenogastrales from Gasteromycetes and couple it on to Hymenomycetes with a necessary re-definition of the characteristics of that group. However, most mycologists will probably, for a long time to come, want to retain the traditional grouping into Hymenomycetes, as at present defined, and Gasteromycetes, and be content to recognize that these are essentially *biological* rather than *natural* taxa.

Assuming the validity of the connecting lines between Hymenomycetes and Hymenogastrales, there remains the recurrent question in a case of this kind of the direction evolution has taken. There are differing views, and some eminent taxonomists support the thesis that Hymenomycetes have been derived from gasteromycete ancestors (Singer and Smith 1960). If this is correct the evidence would suggest that several separate lines must have been involved and, therefore, the basidium as a spore-gun must have developed again and again from non-explosive basidia. This seems extremely difficult to envisage. It is much easier to imagine that the types of basidia found in Gasteromycetes have arisen by degeneration from hymenomycete ancestors. In an almost complete absence of a fossil record, anything like certainty is impossible in dealing with the phylogeny of fungi, but the view is adopted here that Gasteromycetes have, in all probability, been derived from hymenomycete stock.

Having lost the delicate hymenomycete equipment of spore liberation, it seems that Gasteromycetes have been forced to develop methods of dispersal along new and original lines. Nature has tried again. Gasteromycetes may best be understood as a remarkable series of experiments in spore liberation.

It may be legitimate to speculate on the possible causes of this retreat from the hymenomycete condition and, perhaps, the answer is an ecological one. Generally speaking, Hymenomycetes are little adapted to xerophytic conditions. The great majority of fleshy fungi are very sensitive to dryness. Even those leathery lignicolous forms capable of enduring drought can, with few exceptions, continue to liberate spores only under humid conditions. On the other hand, Gasteromycetes reach their fullest development in warm, dry parts of the world and mostly the spore

dispersal mechanisms do not show such dependence on sustained dampness of the air.

There are five major kinds of dispersal in Gasteromycetes: dry-spore
types dispersed by wind, slime-spore fungi spread by insects, hypogeal
forms relying on rodents for dispersal, splash-cup fungi, and the catapult
of *Sphaerobolus*.

The most familiar dry-spore mechanism is to be found in species of
the puff-ball *Lycoperdon* (Fig. 8.2). At maturity the fertile region, or
gleba, is converted into a capillitium of dry springy hyphae, derived from
the trama, saturated, as a powder-puff is loaded with powder, with dry
spores. In *L. perlatum* the capillitium consists of long, thick-walled,
almost wirey, threads about 8 μm wide but often over 1 cm long and very
sparingly branched, or even unbranched. One group of these forms a
central basal tuft in the capsule, the remainder converging inwards from
the peridial wall. This is thin, papery, and unwettable. Eventually, the
capsule opens by an apical ostiole. Wind blowing across this may suck
out spores, but essentially the whole structure seems to be a bellows
mechanism operated by falling water-drops (Gregory 1949). Large
drops several millimetres in diameter from heavy showers of rain or
dripping from trees are involved. A drop striking the unwettable peridium
momentarily depresses it, and a cloud of spores escapes through the
ostiole, the capsule at once resuming its normal form due to the resilience
of the capillitium within.

The same mechanism operates in the earth-stars, for example in
Geastrum triplex. The young onion-shaped sporophore is buried in the
litter of beech leaves with only its pointed end reaching the surface. The
three-layered exoperidium, of which the innermost layer is fleshy,
splits along several lines of weakness extending from the apex towards the
base, thus carving out the 'rays' of the star which eventually bend backwards pushing aside the leaf litter and exposing the capsule with its thin
wall of papery endoperidium. These rays may bend so far that they raise
the capsule significantly and break all connection with the parent mycelium
in the soil. The capsule itself opens by an apical ostiole, which is a much
more specialized structure than in *Lycoperdon*. Within are unbranched
capillitial threads, bound together in fine sheaves, radiating outwards
from a conspicuous columella and inwards from the endoperidium
(Fig. 8.3).

It is interesting to observe an area where *G. triplex* occurs during a
downpour of rain. Being of much the same colour as the beech leaves
among which they grow, the sporophores are sometimes difficult to
locate, but if the rain is heavy they are at once betrayed by the easily
visible puffs of spores arising from them.

Geastrum is a large genus and the species show a considerable range of
structure. In *G. fornicatus*, for example, the outer peridium is divided

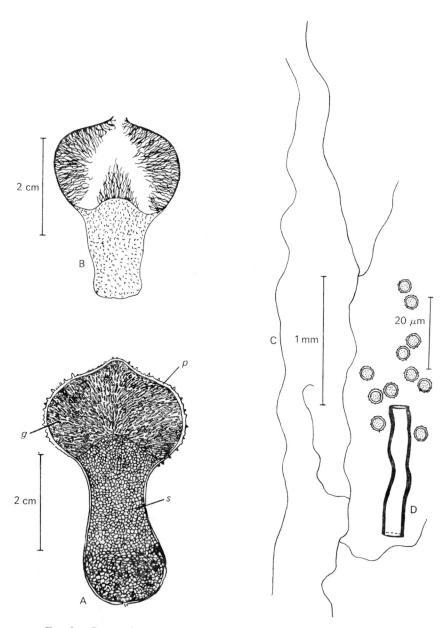

FIG. 8.2. *Lycoperdon perlatum*. (A) fully grown sporophore in longitudinal section showing stipe (*s*), peridium (*p*) and gleba (*g*) in which there are numerous elongated chambers lined by hymenium. (B) mature dry specimen in longitudinal section; the capillitium threads are shown but the spores are omitted. (C) individual capillitium threads. (D) part of a capillitium thread and some spores.

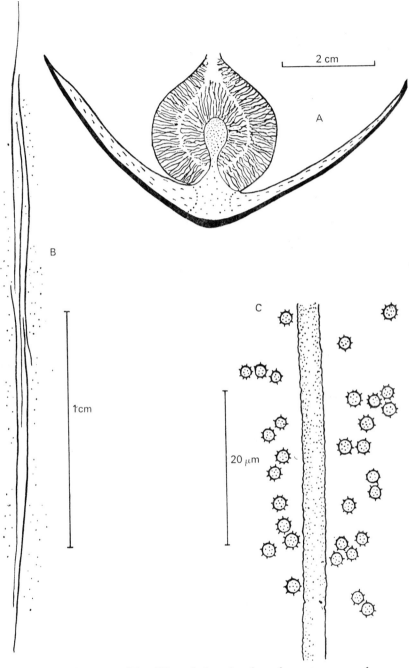

FIG. 8.3. *Geastrum triplex.* (A) vertical section through a mature sporophore showing sheafs of capillitium threads attached to the columella and to the inner peridium (capsule wall). (B) single sheaf of capillitium threads (somewhat separated) with some associated spores. (C) part of a capillitium thread with some spores.

into an inner and an outer part so that at first there are two stellate cups, one inside the other. The inner eventually turns gradually inside out raising the capsule, with its thin wall of inner peridial tissue, into a position more favourable, perhaps, if spores are to be liberated by the action of wind alone, but conferring little advantage if raindrops are the essential agents of spore release (Fig. 8.4).

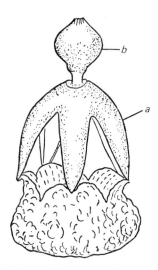

FIG. 8.4. *Geastrum fornicatum.* Ripe sporophore; the inner cup (*a*) of the peridium has turned inside out raising the spore-containing capsule (*b*).

It seems that species of the genus *Tulostoma*, although taxonomically rather remote from both *Lycoperdon* and *Geastrum*, behave in just the same way. *Tulostoma* spp. are, however, essentially desert fungi and the puffing mechanism may be operated by wind-blown sand more often than by raindrops.

Bovista plumbea, a common British pasture species, is very much like a puff-ball (Fig. 8.5). The exoperidium breaks down to a greater or lesser extent leaving the papery endoperidium as the essential capsule wall. The ripe sporophore has a definite apical mouth through which puffs of spores may escape under the action of bombarding raindrops. The capillitium is, however, very different from that of *Lycoperdon*. It consists of a mass of separate units, each a three-dimensional system of branched, stiff hyphae diverging from a central point, and each about 1 mm across. The whole mass of these tiny units saturated with the dry, tailed basidiospores, characteristic of *Bovista*, forms a powder-puff as effective as that of *Lycoperdon* or *Geastrum*. Often the sporophore, surrounded solely

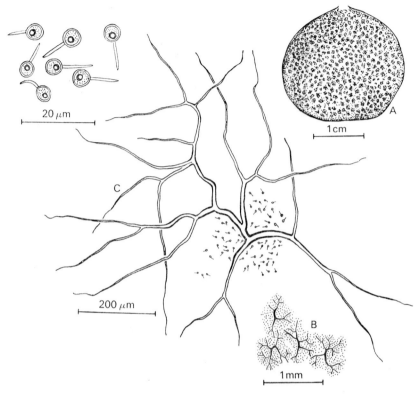

FIG. 8.5. *Bovista plumbea*. (A) Loose capsule surrounded only by papery inner peridium and containing 'crumbs' of dry gleba. (B) four 'crumbs' each with a skeleton consisting of a capillitium element. (C) single capillitium element and a very few associated spores. (D) seven spores.

by the endoperidium, breaks free and, being very light, is easily trundled by the wind, scattering spores as it bounces along without any need of rain to activate its bellows mechanism.

Essentially the same mechanism operated by big raindrops occurs in the large and common slime mould, *Lycogala epidendrum*, the sporocarps of which resemble small puff-balls. A falling drop striking a mature specimen causes it to rupture with the production of an ostiole through which puffs of spores escape when the sporocarp is struck by further drops (Dixon 1963).

The giant puff-balls, included in the genus *Calvatia*, do not have the bellows mechanism of spore liberation. Thus when fully mature the sporophore of *Calvatia caelata* (Fig. 8.6) opens widely to expose a great mass of very dry powdery spores kept loose and uncompacted by a branched system of capillitium threads. Spores are freely blown away by wind. Further, as spores are removed from the exposed surface of

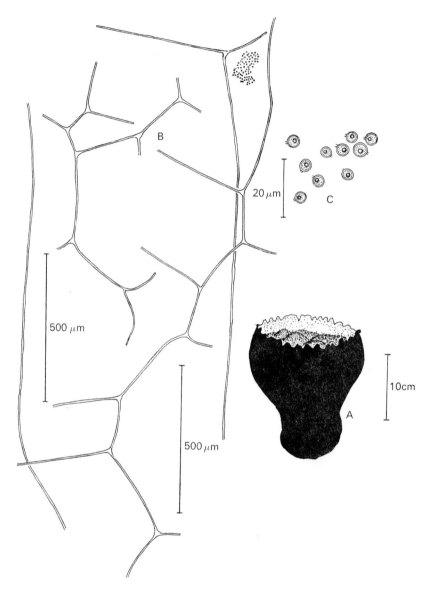

FIG. 8.6. *Calvatia caelata.* (A) dry open sporophore with spore-mass exposed. (B) capillitium threads and a few spores; most of the tips represent naturally broken ends. (C) spores.

the spore mass, the capillitium, which is extremely brittle, is also blown away and thus does not interfere with the liberation of deeper layers of spores.

A very remarkable dry-spore mechanism is found in *Podaxis*. This fungus is of widespread occurrence in warmer countries. The taxonomy of the genus is in a confused state. It has been suggested that only a single, rather polymorphic species is involved, but it seems unlikely that this is the true position. In Ghana, the region in which *Podaxis* is known to the author, there are two clearly defined species. The discussion below is based on the taller species, with a narrower cap and smaller spores, that grows associated with large termite nests. For this the name *P. pistillaris* is provisionally used. *Podaxis*, having a distinct stipe and cap, has a resemblance to certain agarics (Fig. 8.7). In general form it shows a striking agreement with the ink-cap *Coprinus comatus* as that fungus appears at the start of spore liberation. However, the texture is entirely different. The stipe of *Podaxis* is firm, rigid, and almost woody, and the cap tissue forms a dry leathery peridium. The gleba is developed between the stipe, in the cap region, and the peridium. At maturity this gleba becomes a mass of dark brown powdery spores penetrated by very numerous coiled capillitium threads that arise from the stipe and pass outwards to meet the peridium without actually being attached to it. These capillitium strands are often 2–3 cm long, completely unbranched and apparently non-septate. Some of them undergo a spiral splitting giving almost the impression of spirally thickened elaters. When the sporophore is fully ripe the peridium separates slightly from the stipe at the base of the cap and spores sift slowly out. This can happen under the influence of wind and probably also in heavy rain.

Sreeramulu and Seshavataram (1962), working in India, have found that spores of *Podaxis* contributed an abundant element to the air spora there. The mean diurnal periodicity curve showed a very definite maximum around midnight. What determines this is by no means clear but rainfall does not seem to be involved.

Another interesting genus of dry-spore Gasteromycetes is *Battarraea*. The best-known species is *B. phalloides*, but even that would seem to be rare. A large species is illustrated in Fig. 8.8. The mature, expanded fruit-body consists of a dome-like cap borne on a tall and very stiff stalk arising from within a cup-like volva. The ripe glebal mass is at first covered by a thin endoperidium but this ruptures in a rather irregular manner exposing a dry spore-mass held among rather stiff capillitium strands. The chief interest from the point of view of dispersal is that inter-mixed with the spores are much elongated cells with spiral thickening showing a striking resemblance to the elaters of liverworts. Whether by violent movements on drying they play any part in the active liberation of spores is unknown. When elaters from a museum specimen were allowed

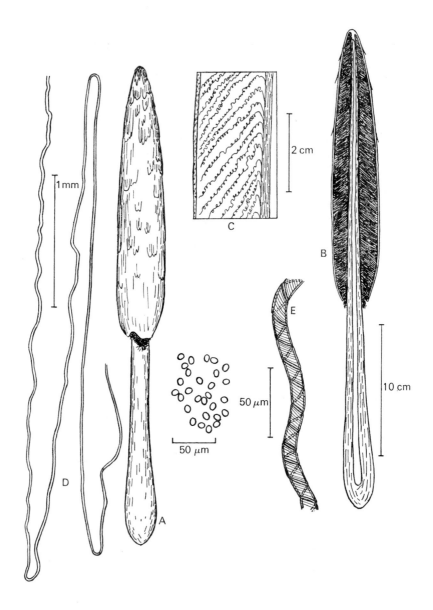

Fig. 8.7. *Podaxis pistillaris.* (A) mature sporophore. (B) same in longitudinal section; spore-mass black. (C) small part of B more highly magnified to show attachment of the unbranched capillitium threads to the stipe and their extension almost to the peridium; spores omitted. (D) part of a capillitium thread bent at three points to fit page. (E) smaller part of (D) more highly magnified, and spores.

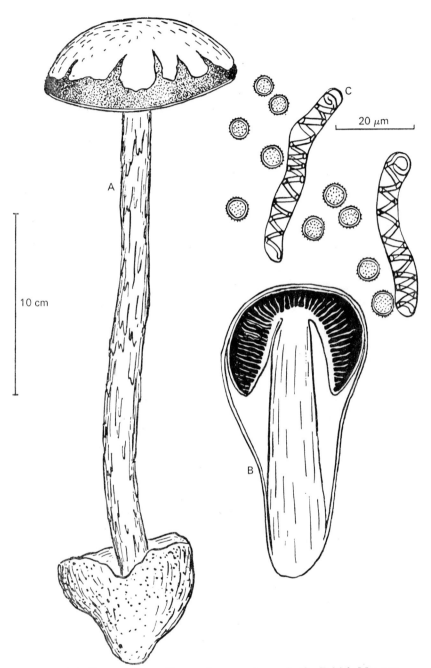

20 μm

10 cm

A

B

C

FIG. 8.8. *Battarraea* sp. (Drawn from specimens in the British Museum (Nat. Hist.) collected in Cyprus). (A) sporophore with woody stipe arising from buried volva; inner peridium peeling off to expose the gleba. (B) longitudinal section of a young specimen; coarse stiff hyphal strands (white) extend outwards through the gleba (black). (C) spores and elaters from the ripe gleba.

to absorb water and then watched during drying no vigorous movements were observed, but this may not have given a true indication of what happens in fresh material.

A feature of many dry-spored Gasteromycetes is the presence of a capillitium that seems to hold the dry spores in a spongy mass from which they can readily, but gradually, be set free. It is interesting to recall that in other organisms such capillitial development has also occurred, for example in certain liverworts (e.g. *Pellia*) and notably in many Myceto-zoa. Further, there are a few scattered examples in other true fungi. Gregory and Waller (1951) have described in *Cryptostroma corticale*, which causes the sooty bark disease of sycamore (*Acer pseudoplatanus*), long capillitial threads that may help to prevent the immediate and mass release of exposed spores. Again, in the smut *Farysia olivacea*, which attacks *Carex riparia*, masses of dry brand-spores replace the ovary and mixed with these are straw-coloured hyphal threads that may be 0·5 cm long. No doubt the result is a more gradual liberation of spores and, perhaps, only by winds that have attained a certain strength. One of the most remarkable examples is the aecium of the rust *Elateraecium salacicola* (Thirumalachar, Kern, and Patil 1966). In this the basal cells abstrict not only aeciospores, but also hyphal cells that become up to 500 μm in length. These are united in a network in which the aeciospores are emeshed. From these woolly masses a centimetre or more in length, hanging down from the diseased leaves of the host (*Salacia prinoides*), the spores may gradually be blown away (Fig. 8.9).

A very familiar example of a dry-spore gasteromycete is the earth-ball, *Scleroderma aurantium* (Fig. 8.10). In this the gleba at maturity becomes a dry, earthy mass of spores not permeated by a significant capillitium. The thickish peridium ruptures in an irregular manner freely exposing the rather large spores which, apparently, are blown away by wind. As a spore-liberating mechanism the earth-ball seems crude indeed by comparison with *Lycoperdon* and *Geastrum*. Nevertheless, dispersal in this fungus seems to be highly effective since it is one of the commonest British species found in great numbers on sandy soil on heaths and in woods.

A remarkable ally of *Scleroderma* is *Pisolithus arrizus*, a rather common fungus in eucalyptus woods in warmer, drier regions of the world such as South Africa, Israel, and Australia (Fig. 8.11). It appears to have mycorrhizal relationship with the trees. When mature the peridium in the apical region of the sporophore breaks down exposing a mass of dry spores unmixed with capillitium as in *Scleroderma*. However, although the upper part may be bone dry, the lower regions of the fruit-body are damp and almost water-soaked. Aqueous drops may exude on the surface in spite of the very dry conditions of the surrounding soil. Further, there is a gradient of ripening. The spores develop in small 'peridiola',

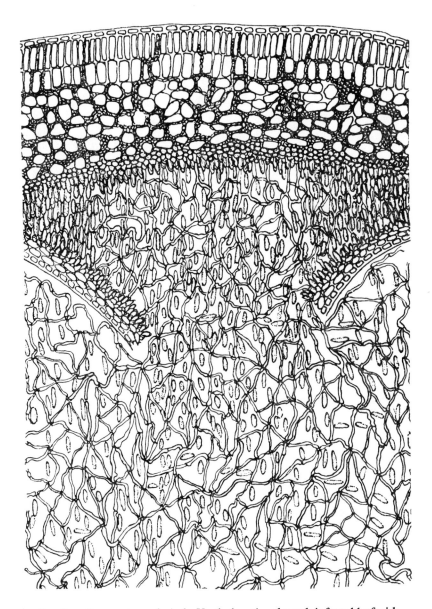

Fig. 8.9. *Elateraecium salacicola.* Vertical section through infected leaf with an aecium. Aeciospores are emeshed in the capillitium, only a small part of which is shown. × 200. After Thirumalachar *et al.* (1966).

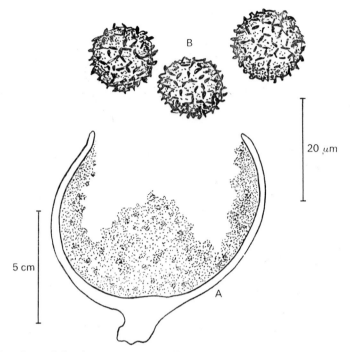

FIG. 8.10. *Scleroderma aurantium*. (A) longitudinal section of a mature opened sporophore containing the earthy spore-mass (gleba). (B) individual spores.

the thin walls of which eventually breakdown so that a general mass of spores is produced. Tracing downward from the ripe apex and following a gradient of increasing water-content, all stages in peridiolar development are encountered. It seems that a single sporophore may last for a long time, possibly for years, fresh crops of spores continually ripening from below, a condition of perennial activity recalling that of *Ganoderma*.

We may now pass to another family of Gasteromycetes: Hymenogastraceae. These are essentially subterranean fungi. Quite a number of the higher fungi have, as it were, gone to earth. Truffles, probably derived from ancestors of the cup-fungus type, are familiar examples. Indeed, within the great group of Ascomycetes comparative studies of structure and development strongly suggest that the hypogeal habit has been evolved along several distinct lines of descent. In the same way, in Gasteromycetes hypogeal forms have arisen and, as in Ascomycetes, the evidence indicates a number of independent evolutionary lines.

It would appear that the hypogeal fungi generally have the same dispersal story. Sporophores when ripe give out a smell detectable by rodents which grub up and eat them. The spores pass through the alimentary canal and are deposited unharmed in the dung. It has to be admitted,

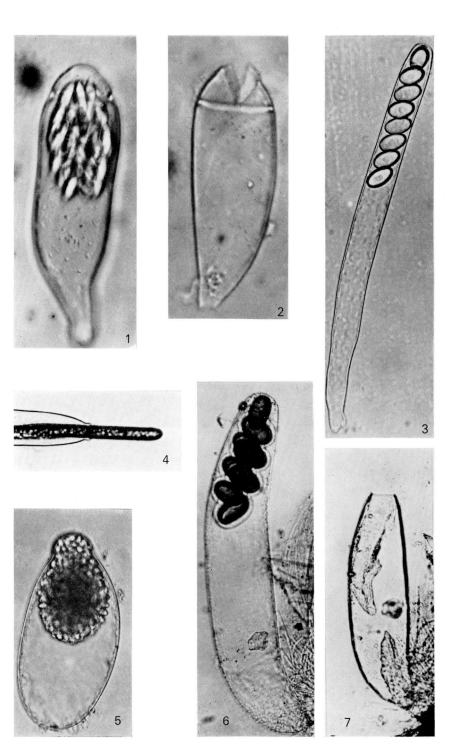

PLATE I. Asci. **1, 2,** *Ascozonus woolhopensis,* a ripe ascus and a discharged ascus; note ring of thickening near top of ascus. × 700. **3,** *Pyronema ompha-lodes,* ripe ascus. × 500. **4,** *Geoglossum* sp., ascospore escaping from the ascus. × 260. **5,** *Rhyparobius nanus,* multispored ascus. × 500. **6, 7,** *Dasyobolus immersus,* mature ascus with purple spores surrounded by mucilage sheaths and the same ascus a few moments later after discharge. × 140.

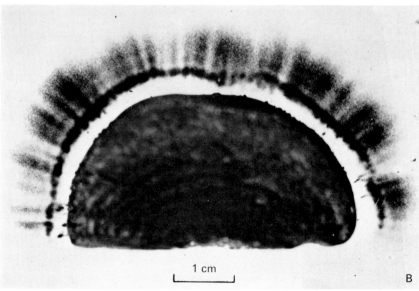

1 cm

PLATE II. *Daldinia concentrica*. **A,** rather small stromata on dead branch of ash (*Fraxinus*). **B,** spore deposit, showing zoning, accumulated overnight around a thick median slice of a largish stroma lying horizontally on a glass plate.

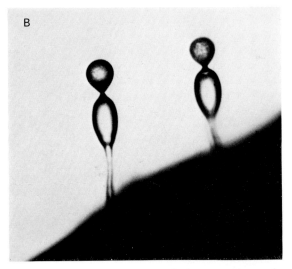

PLATE III. A, *Pilobolus kleinii*, sporangiophores on horse dung. ×12.
B, *Basidiobolus ranarum*, conidiophores on the excrement of frogs. ×260.

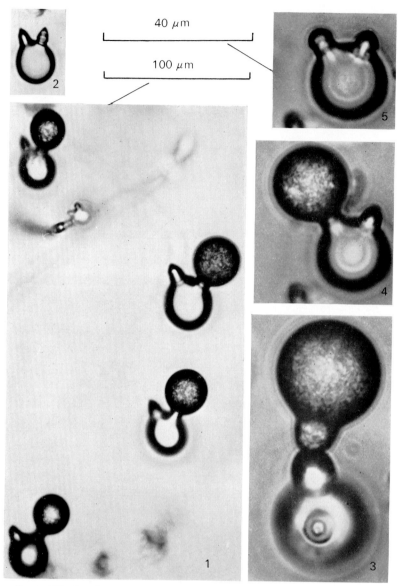

40 μm

100 μm

PLATE IV. *Conidiobolus coronatus.* Discharged conidia on water agar, 3 h after discharge, which have germinated to produce secondary conidia. 1, Four conidia each with a ripe secondary conidium (produced towards the light); the projection to the left of each being the everted papilla of the orginal spore produced at the moment of discharge. 2, Topmost spore of the four in the previous figure when the secondary spore has been discharged (note the fortuitous resemblance of the vacated conidiophore and the everted papilla of the parent spore). 3, A secondary conidium where the papilla has become everted but not vigorously enough to discharge the spore; the everted papilla of the primary spore is seen in surface view. 4, An empty primary conidium with its papilla on right and with a ripe secondary conidium on the left. 5, The same a few minutes later.

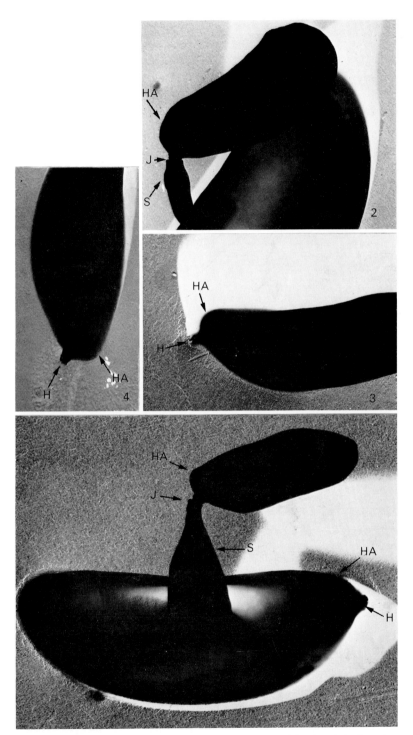

PLATE V. *Sporobolomyces roseus*. Electron micrographs of dried and shadowed specimens. **1, 2,** Discharged ballistospores that have germinated by a sterigma (S) to produce a secondary ballistospore. **3, 4,** Freshly discharged ballistospores. H, hilum; HA, hilar appendix; J, junction of spore and sterigma.
× 14 000.

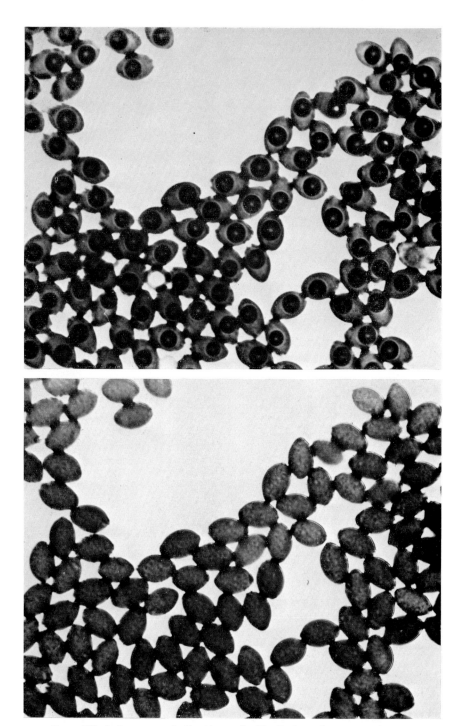

PLATE VI. *Podospora curvicolla*. Above: freshly discharged spores caught on a dry slide photographed 20 s after flooding with water; each contains a gas bubble. Below: the same spores 60 s later; the gas phase has disappeared. × 630.

PLATE VII. *Sphaerobolus stellatus*. Group of fruit-bodies as seen under a simple lens. In some the dark glebal-mass is still undischarged and lies submerged in the 'lubricating fluid' within the inner cup of the fruit-body. In others discharge has occurred and the inner cup has turned inside out and now has the appearance of a pearl. The two bright spots appearing in the undischarged fruit-body are images of lamps used in the photography.

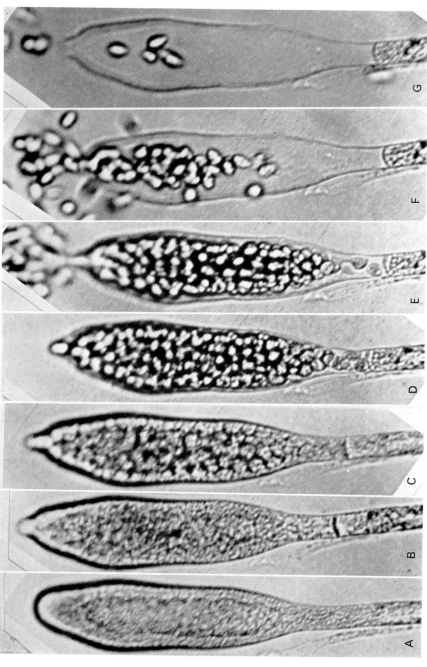

PLATE VIII. *Saprolegnia ferax*. Maturation of zoosporangium and release of zoospores. **A**, swollen hyphal tip with large central vacuole but no delimiting basal cross-wall. **B**, basal cross-wall formed and apical papilla developed. **C** and **D**, zoospore delimitation. **E**, **F**, and **G**, stages in spore escape. Only a few seconds separate stage **D** from stage **G**. × approx. 500. Stills from a film by Dr. D. Greenwood.

however, that this story is largely based on inference, and there is very little observational or experimental basis for it. Studies on the essential dispersal story of the hypogeal fungi would be very welcome.

One sharply characterized order of Gasteromycetes, Phallales, has fruit-bodies obviously related to insect dispersal of spores. There are some twenty genera and their richest development is in Australasia and tropical countries.

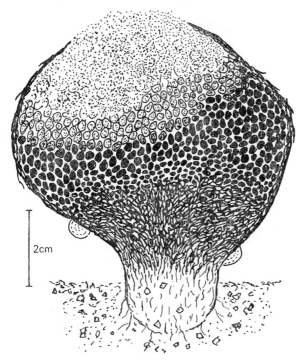

Fig. 8.11. *Pisolithus arrizus.* Vertical section of fruitbody. Peridium has ruptured and mass of dry spores is exposed. Below there are pale brown peridiola filled with ripe spores, and beneath these again developing water-soaked peridiola. Liquid droplets exuding near base. Soil has angular fragments some of which are included in the indefinite stipe.

The sub-mature fruit-body in most phalloids is like a soft egg (Fig. 8.12) formed just below the soil or leaf-litter level at the end of a rhizomorph traceable back to buried wood. Finally, in most genera a spongy stalk within the 'egg' elongates with remarkable speed carrying the slimy spore-mass above the ground to a height of several inches. In *Clathrus* and its allies there is no stalk, but the spore-slime coats a more or less spherical open network that quickly expands soon after the 'egg' ruptures.

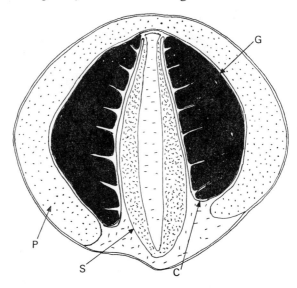

Fig. 8.12. *Phallus impudicus*. Fully grown 'egg' in longitudinal section. (P) gelatinous middle peridium; (S) stipe; (C) cap; (G) gleba. Natural size.

In all species the exposed spore-slime contains abundant sugar and emits a strong and unpleasant odour. The basidiospores themselves are minute and smooth-walled. This is a type of spore encountered repeatedly in fungi where insect dispersal is the rule.

It is of some interest to compare the two commonest British phalloids: *Phallus impudicus* (stinkhorn) and *Mutinus caninus* (dog's stinkhorn). *P. impudicus* has a strong odour detectable at a distance of many yards, but *M. caninus* has to be held quite close to be smelt. Stinkhorns usually expand from the 'eggs' in the forenoon. In *P. impudicus* the cap is seen crowded with flies and bluebottles, and by evening it is normally left white and spore-free (Fig. 8.13). On the other hand, in *Mutinus* flies are not very often seen eating the slime and very frequently collapsed specimens can be collected with the spore-load undiminished. It is tempting to suggest that this difference in these two stinkhorns is connected with the difference in the intensity of their smell.

It is natural to try and interpret the structure of phalloids in terms of possible importance in dispersal. What may be the biological significance of the remarkable network crinoline of *Dictyophora* (Fig. 8.14B) is by no means clear, but the scarlet sterile rays of *Aseroë* (Fig. 8.14A), surrounding the central mass of spore-slime, are very suggestive of the brilliant petals of an entomophilous flower, and there is little doubt that they have the same function.

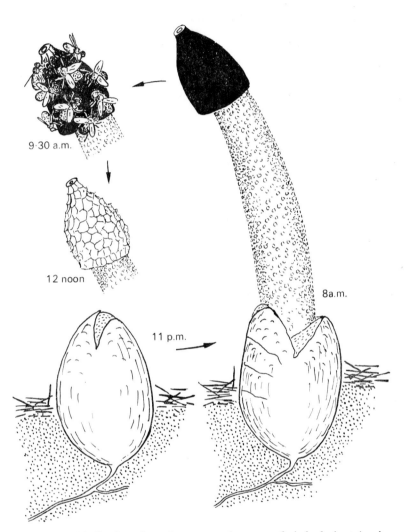

F IG. 8.13. *Phallus impudicus.* At 11 p.m. the young fruit-body is at 'egg' stage and the outer papery layer of the peridium has torn exposing the jelly of the middle peridium. Next morning at 8 a.m. the stipe has elongated carrying up the cap with the spore-slime and leaving the peridium as a volva around the base. By 9.30 a.m. the slime is giving out a strong smell and has attracted flies; by noon all the spore-slime has been removed.

The dispersal story in Phallales is somewhat unsatisfactory because of lack of knowledge of the fate of the spores. Germination has not yet been observed. Spores that have passed through the alimentary tract of flies appear quite unaltered, but, as with those taken directly from the spori-ferous slime, they fail to germinate under laboratory conditions. Indeed, in general, gasteromycete spores are difficult to germinate (Bulmer and Bencke 1964). It would, however, seem unlikely that the spores of stink-horns are no longer capable of germination. The probability is that the conditions in nature that induce germination have not yet been identified. *Phallus impudicus* is a very common species with a considerable geographic range. Its mechanism of dispersal may, therefore, be assumed to be efficient and there does not seem to be any subsidiary means of spread apart from the basidiospores.

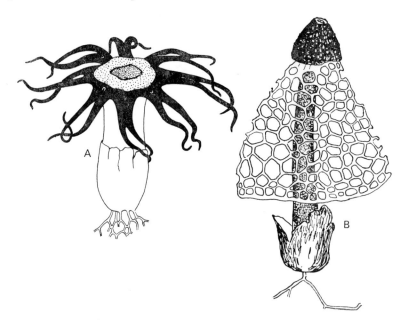

Fig. 8.14. (A) *Aseroë rubra :* the central stippled area is the spore-containing slime, the bifid sterile rays are crimson; after Massee. (B) *Dictyophora indusiata :* sporophore showing the net-like 'crinoline' below the cap with its spore-slime; for simplicity only the front half of the 'crinoline' is shown.

Perhaps the most extraordinary specialization in connection with dispersal is to be seen in Nidulariales, but the mechanisms in its two families, Nidulariaceae and Sphaerobolaceae, are so different that they must be considered separately.

The family Nidulariaceae embraces a number of genera in all of which the sporophore contains within it a small number of hard, almost seed-like, packets of spores known as peridiola. The sporophore opens at maturity to expose the peridiola which are liberated for the most part by rain-splash. The significance of the form of the fruit-body was appreciated by Buller near the end of his life, and the story was later developed in detail by Brodie (1951, 1956) who recognized a series: *Nidularia*, *Nidula*, *Crucibulum*, and *Cyathus* in order of increasing splash-cup efficiency.

Cyathus striatus (Fig. 8.15) is a fairly common and widely distributed species found usually on rotting wood. It was illustrated beautifully by the Tulasne brothers over a hundred years ago and Brodie has maintained their traditions in his fine drawings of structure in relation to function in this species.

When the sporophore is fully mature it is an open vase about 1·5 cm high with a circular mouth roughly 1 cm across. It is firmly emplaced on the woody substratum and contains a dozen or so peridiola, each a hard seed-like structure, discoid, and 2–3 mm in diameter. In section the peridiolum is seen to have a wall of several layers, including a very hard one, and there is a central flattened cavity lined by hymenium. However, at maturity the basidia are disorganized and the centre is filled with a mass of spores. Each peridiolum is attached by a short stalk, 2–3 mm long, to the inner wall of the peridial vase. The stalk is clearly differentiated into a lower part (sheath) and an upper part (purse). The purse is a delicate, hollow, tubular structure closed below, and coiled up within is a rope of hyphae (funiculus) associated like the strands of a wire cable. The upper end of this is firmly attached to the peridiolum, but the lower, free end is frayed out to form the hapteron, a mass of very sticky hyphae. The purse is connected with the sheath by a sheaf of hyphae forming the 'middle piece'. This can be demonstrated if a peridiolum is carefully pulled away from its attachment.

A large drop of water falling into the open sporophore is broken up and reflected from the splash-cup as a number of separate droplets some of which may carry peridiola to a distance of a metre or two. Under the impact of the drop the peridiolum with its attached funiculus breaks free from its stalk by rupture of the purse, and when this happens the hyphae of the hapteron spread out almost explosively. Probably the funiculus is still coiled in the reflected droplet as it travels through the air. If the droplet strikes such an object as a fine stem, the hapteron adheres strongly and, although the peridiolum may be carried further by its momentum and the funiculus be stretched to a length of 5–10 cm, the peridiolum is finally arrested by its efficient tether. Sometimes this may lead to the funiculus becoming coiled several times around a narrow stem or the peridiolum may simply be left hanging from such a horizontal object as a leaf.

The subsequent fate of the peridiolum is not known with any certainty. Falling on rotten wood it may germinate directly to produce a

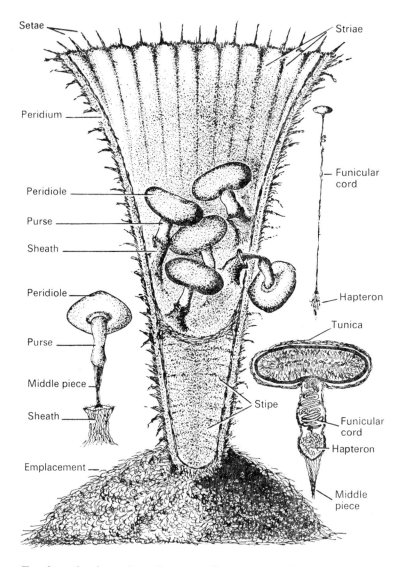

FIG. 8.15. *Cyathus striatus.* Structure of mature sporophore. After Brodie (1951).

dikaryotic mycelium without the contained spores being involved at all. It is to be noted, however, that quite a number of Nidulariaceae are coprophilous. For example *Cyathus stercoreus* is a common species on horse or cow dung, or on heavily manured ground. In coprophilous types it may well be that, as with other dung fungi, the peridiola are eaten with the herbage and germination occurs after passing through the alimentary tract. Brodie has, indeed, observed that the spores of *C. stercoreus* germinate best following mild heat treatment such as might be experienced in passage through an animal.

Cyathus striatus seems highly specialized in connection with splash dispersal. Even the striae lining the vase and the setae around its rim look as if they have significance in connection with droplet reflection.

As we descend Brodie's suggested series specialization seems to decrease. There is still a well-developed funiculus in *Crucibulum*, but the form of the splash-cup does not appear so perfect as that of *Cyathus*. In *Nidula* the spash-cup is still there but the rather small peridiola are devoid of funiculus. Instead each is sticky over its whole surface. In *Nidularia* the spherical sporophore ruptures in an irregular manner and, although splash dispersal may well occur, the structure can hardly be called a splash-cup.

Sphaerobolus is classified in Sphaerobolaceae, a family usually included in Nidulariales. The taxonomy of the genus is difficult, but probably the majority of collections can be assigned to *S. stellatus*, a common species found in most parts of the world growing on very rotten wood or old dung pads. Further, unlike other Gasteromycetes, it grows and fruits readily in pure culture on a suitable medium such as oatmeal agar, provided the light intensity is sufficient and the temperature is below 25°C. In maintaining the fungus in a fruiting condition it is best to start new cultures from discharged glebal bodies (peridiola).

S. stellatus (Fig. 8.16 and Plate VII) is one of the smallest of Gasteromycetes with a fruit-body only about 2 mm in diameter. From this at maturity a spherical peridiolum, or glebal mass, 1 mm across is catapulted to a distance of several metres. A detailed account is given by Buller (1933).

The nearly ripe fruit-body is spherical and anchored to the substratum by a cottony mycelium. In vertical section it is seen to consist of a spore-containing gleba surrounded by a peridium of six histologically distinct layers. When mature it splits along a number of lines radiating from the apex, thus carving out four to eight teeth of peridial tissue which bend outwards exposing the spherical glebal mass. At the same time the peridium separates into two cups, one fitting inside the other and each with a toothed margin. The cups remain in contact only at the tips of the teeth. The outer cup is composed of the three outermost layers of the peridium,

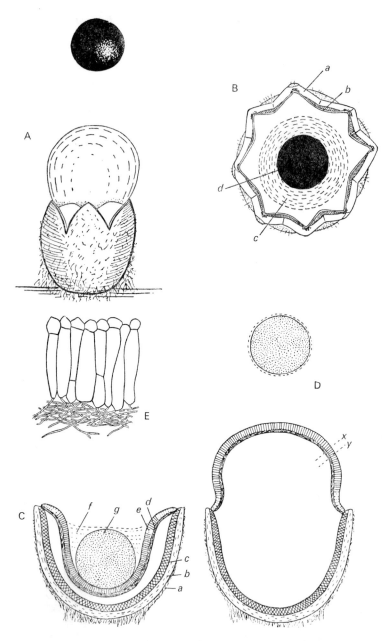

FIG. 8.16. *Sphaerobolus stellatus*. (A) Sporophore at instant of discharge;
the inner cup has turned inside out exposing its wet, shining surface. (B)
looking directly into an opened sporophore before discharge: (*a*) outer cup;
(*b*) inner cup; (*c*) lubricating fluid in which the glebal mass (*d*) is submerged.
(C) vertical section of opened sporophore. (*a, b, c*) tissue layers of outer cup;
(*d* and *e*) tissue layers of inner cup; (*f*) lubricating fluid; (*g*) glebal mass.
(D) same at moment of discharge. (E) part of inner cup (between lines *x*
and *y* in (D)) highly magnified. (A–D) × 15.

and the inner cup of two of the remaining layers. The sixth layer, which was in immediate contact with the gleba, has largely autolysed at this stage to produce a fluid, so that the glebal mass fits loosely in the inner cup just submerged by liquid.

Contributing largely to the inner wall of the inner cup is a palisade tissue of relatively big cells, but the outer wall is composed of fine interwoven tangential hyphae. By absorption of water the inner surface of the cup tends to increase. Thus strains are set up which are suddenly and violently released by the inner cup turning inside out, thereby catapulting the glebal mass to a distance of up to 7 m.

Walker and Andersen (1925) pointed out that before the sporophore opens the palisade cells of the peridial layer are rich in glycogen. As in so many structures concerned with violent discharge in fungi, this disappears in the later stages of ripening and is presumably converted to sugar with a consequent increase in osmotic pressure leading to a build-up of turgor. Walker considered that this glycogen was converted to maltose. However, more recent study by Engel and Schneider (1963), using paper chromatography, indicates that glucose is the sugar produced, the amount in the tissues of the inner cup increasing rapidly from the first appearance of stellate splitting of the sporophore until it is wide open and ready to discharge its glebal mass.

The soft sticky glebal mass has a thin brown wall composed of disorganized peridial cells. It contains very numerous minute spores and a much smaller number of somewhat larger dikaryotic cells or gemmae. On a suitable substratum the glebal mass germinates immediately producing dikaryotic mycelium, apparently derived from the gemmae. The basidiospores tend to remain dormant. However, Walker has succeeded in inducing their germination in water containing a little pepsin.

Sphaerobolus is both a lignicolous and a coprophilous fungus. As pointed out by Buller (1933), it seems to have all the features of specialized coprophilous fungi shown also by *Pilobolus* spp., *Dasyobolus immersus*, and *Podospora* spp.: there is a large and strongly adhesive spore-mass discharged sufficiently far to reach the grass around the dung; the spores are screened from light after discharge; and the spore-gun is aimed by positive phototropism.

Buller (1933) reviewed the earlier work on phototropism in *Sphaerobolus* and added some observations of his own. He considered that the action of light in orientating the sporophore is limited to the very young stage. However, it has now been shown (Nawaz 1967) that throughout the development of the sporophore there is directional response to light and this may even occur during the last few hours when stellate opening has already occurred.

When *Sphaerobolus* occurs on dung there is little doubt that it has found its way there from discharged glebal masses, on grass eaten by the animal, by way of the alimentary canal. Thus it behaves like most of the other members of the coprophilous flora.

The picture of the wonderful range of dispersal mechanisms in Gasteromycetes has been given further interest recently by the discovery of a remarkable marine gasteromycete (*Nia vibrissa*) with spores beautifully adapted to its aquatic life. This extraordinary fungus is, however, best considered in the final chapter.

9. BLOW-OFF, SPLASH-OFF, AND SHAKE-OFF

THERE are very many fungi in which spore liberation is a passive process in the sense that outside kinetic energy is responsible for spore liberation. This chapter deals with this important kind of release which, perhaps because it is unspectacular, has seldom been the subject of careful study.

It is difficult to decide if passive or active liberation mechanisms are the most usual in fungi, but in this connection it is of interest to examine a particularly well-documented case. Lacey (1962) using a Hirst spore-trap analysed the air spora 0·5 m above ground level at two sites 450 m apart in a rural area west of London during summer (May to September). The catches from the two places were in general similar. Table 9.1 gives, in a

TABLE 9.1

Fungal spores in summer air, Silwood Park		
	Percentage total catch	
	Site S	Site M
Ascospores	14	7·5
Ballistospores	56	47·5
(Mirror-image yeasts)	(35)	(28·5)
Conidia	17	32
(*Cladosporium*)	(15·5)	(28·5)
Other spores	13	13

Note. Most conidia belong to *Cladesporium* and most ballistospores to mirror-image yeasts.

greatly condensed form, a summary of Lacey's results. In this particular spora the two types of spore that are violently discharged are ballistospores and ascospores; for the remainder liberation is passive. The allocation to these different categories depends to some extent on ease of identification. Ballistospores shed from hymenomycete sporophores and discharged from the sterigmata of 'mirror-image' yeasts (Sporobolomycetaceae) are easily identified because of the characteristic projecting hilar appendix. Ascospores are much less recognizable, and it is quite likely that in the category 'other spores' (Table 9.1) a number of these may be included.

It is clear, however, that in this spora violently discharged spores, accounting for 55–70 per cent of the total, outweighed those liberated passively;
but probably in other instances the reverse is true.

In a large range of dry-spored moulds, which grow mainly as saprophytes on dead vegetation, the spores may be blown off their conidiophores by wind. In many genera (e.g. *Cladosporium, Botrytis, Trichothecium, Aspergillus, Penicillium*) the conidiophore has an erect stalk that
raises the spores slightly (usually 0·1–2 mm) above the substratum, thus
increasing their chances of take-off under the influence of eddies that
may break into the laminar layer of air in which they are immersed for
most of the time. In addition the spores may be shaken off their conidiophores when the substratum is violently agitated by wind or heavy rain.

There has been comparatively little quantitative study of spore liberation from moulds of this kind. Zoberi (1961) used a set-up (Fig. 9.1)

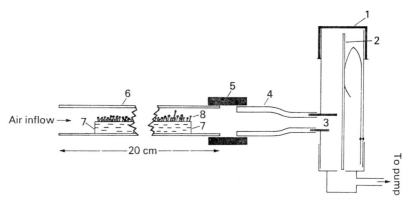

FIG. 9.1. Diagram of impactor unit connected to a culture-tube (half real
size): (1) moulded rubber cap; (2) glass slide; (3) entry slit; (4) steel con-
nector; (5) rubber tube; (6) culture-tube (middle part omitted); (7) agar
strip; (8) aerial growth of fungus.

in which air at controlled speed and humidity was drawn over a culture in
a horizontal glass tube, the liberated spores being then impacted on a
sticky glass slide. A very similar system was used by Pady *et al.* (1969c)
in a study of spore release in *Cladosporium*. Zoberi experimented with
a number of dry-spore moulds and his results for *Trichothecium roseum*
may be taken as fairly representative. As might be expected, the number of
spores set free increased markedly with the rate of airflow. Further, at a
particular speed, spore liberation fell off rapidly with time, most spores
being dislodged with the first blast of air (Fig. 9.2). Again spores were
much more easily blown off their conidiophores when the air was dry
than when it was nearly saturated (Fig. 9.2). The same seems to be true
for *Cladosporium* (Pady *et al.* 1969c).

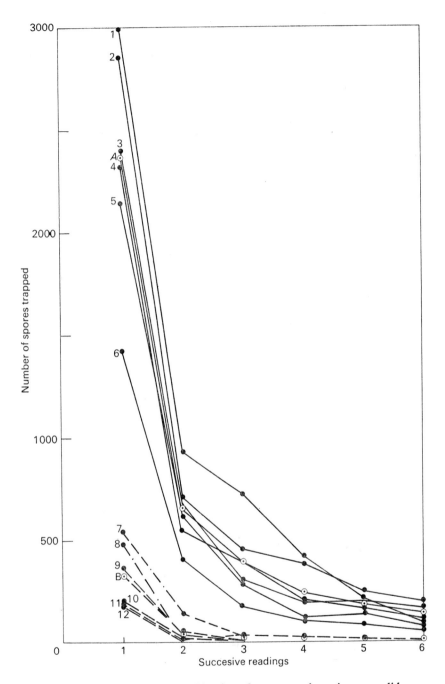

FIG. 9.2. *Trichothecium roseum.* Number of spores caught on impactor slide in six successive 1-min intervals; graphs 1–6 for culture-tubes subjected to dry air-stream (50 per cent r.h.), (*A*) being the mean curve; graphs 7–12 for the tubes subjected to damp air stream (95 per cent r.h.), (*B*) being the mean. Air speed 5 m/s.

A rather similar study was undertaken by Smith (1966) using stem-rust of wheat (*Puccinia graminis tritici*). A diseased leaf was held in midstream in a horizontal tube, acting as a tiny wind-tunnel, through which air in a 'smoothed' (i.e. non-turbulent) condition passed at a controlled speed. A certain minimum was necessary to achieve any liberation of urediospores, but above this the number of spores blown off increased directly with wind velocity (Fig. 9.3). Although the urediospores are

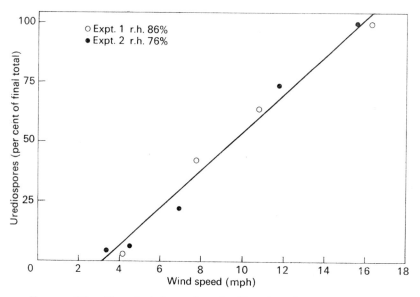

Fig. 9.3. The effect of wind speed on the liberation of urediospores from *P. graminis* pustules.

borne singly on short simple stalks, nearly half the spores trapped were in groups of two or more. This rather suggests that in the urediosori the spores actually separate from their stalks before being in a state to be liberated by wind. Essentially the same point is made by Jarvis (1962) in relation to the liberation of conidia of *Botrytis* from raspberries affected by 'grey mould'. Probably in many dry-spore moulds it is necessary to distinguish between spore-separation and actual take-off.

Blow-off of dry spores is a feature of a number of the larger fungi, especially puff-balls and their allies. It is also the characteristic mode of release in ordinary slime-moulds (Mycetoza). Frequently the spores are presented to the wind as a loose mass supported by a system of capillitial threads. However, this matter has already been discussed in connection with the extraordinary experiments in spore dispersal displayed by Gasteromycetes (Chapter 8).

Although splash has long been recognized as a mechanism of spore liberation (Faulwetter 1916*a, b*) it is only recently that the process has received systematic study (Gregory, Guthrie, and Bunce 1959).

Interest clearly centres mainly on raindrops. These vary from 0·2 to 5 mm in diameter, still larger drops being unstable and breaking into smaller ones. In rain a steady terminal velocity of 7·8 m/s is reached by the largest drops, but for those only 1·0 mm in diameter this is reduced to 1·0 m/s. Secondly, dripping drops may be effective in spore liberation, particularly under trees. Here the kinetic energy of the drop is an important consideration and this will approach its full value for a large drop only if it has a sufficiently long path of fall, of the order of several metres.

When a large drop strikes a thin film of water, it breaks up and is reflected as a centrifugal pattern of droplets which are intimate mixtures of drop and film. This can be shown by adding dyes of contrasted colour to the drop and to its target, and observing the mixed pigmentation of the resulting splashes (Gregory 1952).

Gregory and his co-workers (Gregory *et al.* 1959) have studied splashing, using drops 2–5 mm in diameter falling from a height of 2·9 or 7·4 m onto films of spore-containing water ranging from 0·1 to 1·0 mm in thickness. The spores mostly used were those of *Fusarium solani*, a slime-spore species. A drop 5 mm wide falling 7·4 m onto the thinnest (0·1 mm) of the films produced over 5000 reflected droplets varying in size from 5 to 2400 μm. They were thrown to horizontal distances up to 100 cm and a large proportion contained spores.

As a more natural target Gregory and his associates used a sycamore twig, bearing conidial stromata of the coral-spot fungus *Nectria cinnabarina*, inclined at an angle of 45°. From a single large impact drop, 2000–3000 droplets were produced, mostly less than 100 μm in diameter and all of them carried spores. Over a hundred of the droplets from this single splash were less than 20 μm in diameter and were therefore of a size to become incorporated in the air-spora (Fig. 9.4). Thus, although splashing mostly throws spores to only a short distance, usually less than 0·5 m, it may lead to minor recruitment to the air spora, and, indeed, all extensive lists of fungi from the air include some slime-spore species.

Although the distance of splash is nearly always short, a spore deposited by rain-splash on a branch or a leaf may be resplashed by another large raindrop. Thus in a heavy storm a very few spores may be carried considerable distances from their original source in a series of hops.

So far the situation has been considered in which the spore-load is in the target, not in the incident drop. However, as pointed out by Gregory *et al.* (1959), the reverse condition may also be envisaged in which a drip-drop containing abundant spores falls from a height onto a spore-free surface from which spore-containing droplets are reflected. For example,

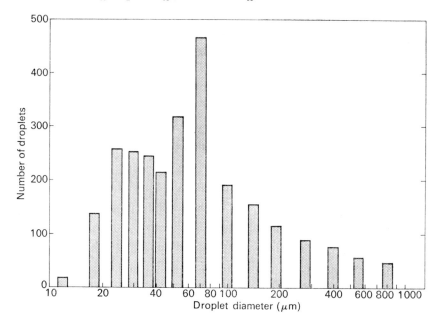

FIG. 9.4. Number of droplets, classified in size-ranges, from one splash by a drop 5 mm in diameter falling 7·4 m on to a wetted twig of *Acer pseudo-platanus* bearing conidial fructifications of *Nectria cinnabarina*. After Gregory, Guthrie, and Bunce 1959.

such a drop might be produced, during a shower, at the end of a downward-inclined twig covered with the pink conidial pustules of coral spot. Some of the features of splash-dispersal are illustrated, very diagrammatically in Fig. 9.5.

It is reasonable to suggest that most slime-spore fungi are dispersed by insects or rain splash or by both. In the stinkhorn fungi (Phallales) and in *Ceratocystis* spp. the slime-spores are largely, if not exclusively, insect dispersed, but in most species of *Fusarium*, *Gleosporium*, and *Colletotrichum* probably splash is of essential significance. In a number of slime-spore fungi the spore-producing region is dry and hard in fine weather, and it is only following wetting, when the mucilage associated with the spores has had time to absorb water and swell, that they are in a condition for effective splash-dispersal.

Detailed field studies of the efficacy of splash in the distribution of spores are few, but the situation in apple trees in connection with the *Gleosporium* rot of the fruit has been considered in some detail by Corke (1966).

Bock (1962) has also studied this question in relation to the dispersal of urediospores of the rust *Hemileia vastatrix* by rain within coffee plantations. He estimated the deposition on healthy leaves of urediospores

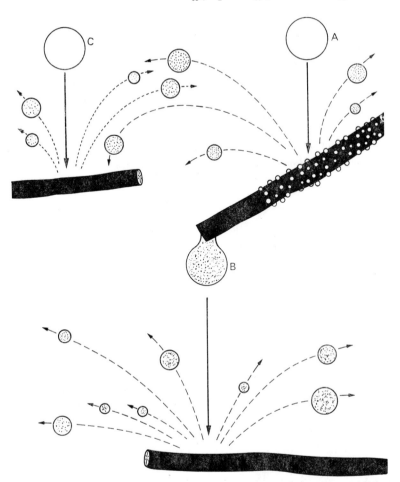

FIG. 9.5. Diagram of splash dispersal from an inclined twig bearing conidial stromata of *Nectria*. Incident paths shown by solid lines; reflected ones by dashed or dotted lines. (A) spore-free drop; (B) drop accumulated at end of twig and rich in spores; (C) spore-free drop involved in 'secondary' dispersal.

derived from diseased foliage on the same trees. Great numbers were deposited but only after heavy showers when the daily rainfall exceeded 0·8 cm. On days of heavy precipitation the number of spores deposited on the leaves was roughly proportional to the intensity of the showers. He considered that wind played little part in the spread of the disease and that splash was the major factor in local dispersal.

The role of rain-splash in the dispersal of the peridiola of birds'-nest fungi (especially *Cyathus*) has already been discussed (Chapter 8).

The importance of rain-splash in dispersal is by no means limited to slime-spore fungi, as illustrated for example by the work of Jarvis on grey mould (*Botrytis cinerea*) on raspberries, but, where these fungi are concerned, the overall mechanical action of rain, in causing violent agitation of the substrata bearing the moulds, is probably of major significance. Nevertheless drops can pick up both slime spores and dry spores. When drops of water become loaded with spores by contact with a spore-bearing surface, their distribution within the drop will depend on their nature. Slime spores, as in *Fusarium* spp. or *Verticillium* spp., will become evenly distributed throughout the drop, but spores that are difficult to wet, like those of *Penicillium* spp., are likely to be limited to the surface. When a drop with slime spores falls on a hydro-phobic surface, say on a leaf, and drains off, it will carry most of its spore-load with it. However, a drop with unwettable spores on its surface will probably leave most of these behind in its wake as it flows over the leaf before dripping off the tip (Davies 1961; Holliday 1969).

Hirst and Stedman (1963), working at Rothamsted, have attempted to analyse the processes involved in the liberation of dry spores by large falling raindrops. They allowed drops that had fallen from a height of 7·6 m to pass through a hole in the top of a small and otherwise closed chamber and strike the spore target. The air in the chamber was sampled by a volumetric impactor trap and a technique was used that allowed impacted spores associated with droplets to be distinguished from those that were deposited dry.

One type of target consisted of a thin layer of dry *Lycopodium* powder exposed on either a slab of steel 3·8 cm thick, and therefore not liable to vibration, or on a celluloid disk resting on short grass that would easily vibrate when struck by a drop. To bombard these targets, drops of water, or glass beads of comparable size, were used. From both targets both types of 'bomb' set dry spores free, but the drops were considerably more efficient than the beads, particularly where the steel block was concerned. Thus the beads liberated comparatively few spores from the block, but a much greater number from the celluloid disk.

The explanation of these results seems to be that when a drop falls on the steel block, the velocity of outward radial flow of the liquid is so great that the air in the laminar boundary layer, in which the spores are con-tained, is momentarily set into violent motion lofting spores. On the other hand, with the bead, since it is not significantly distorted by impact, no comparable disturbance of the air is involved. When, however, the bead falls on the cellulose disk there is considerable vibration of the target leading to substantial spore release. The water drop is less efficient in lofting spores from this target, as compared with the rigid steel surface, probably because it dissipates energy in a momentary depression of the celluloid.

Their work led Hirst and Stedman to distinguish 'tap' and 'puff' in the liberation of dry spores by falling raindrops. In the 'tap' process the spores are *shaken* free, in the 'puff' process they are *blown* off by a local high-speed pressure wave produced at impact. Normally 'tap' and 'puff' act together and their effects cannot be separately distinguished. However, an exception was noted using leaves of hop mildewed by *Pseudoperonospora humuli*. The sporangiophores occur only on the under surface. When such leaves, held in the natural position, were struck with a falling drop Hirst and Stedman observed that 'a small, black, eddying cloud issued from below the impact point'. This is clearly an example of 'tap' on its own.

In addition to experiments with *Lycopodium* powder, Hirst and Stedman used a range of natural fungal targets and found that falling drops liberated spores in such dry-spore forms as *Cladosporium, Alternaria, Epicoccum, Ustilago nuda*, and *Puccinia* spp. (urediospores). For all these the Rothamsted workers have noted temporary increases in the air spora at the onset of rain. Indeed, in general, stem vibration and leaf flutter caused by heavy rain must be very significant factors in spore liberation, but it is to be remembered that the same rain very soon combs the liberated spores out of the air and brings them to earth.

As we have already seen large raindrops also play a part in operating the bellows mechanism of dry-spore liberation in puff-balls (*Lycoperdon*), earth-stars (*Geastrum*), and certain slime moulds (e.g. *Lycogala*).

In considering spore liberation by water drops, attention has been focused on large ones falling with a considerable velocity, but there is also the possibility that the very much smaller droplets of mist, moving mainly in a horizontal direction, may pick up spores and transport them. Glynne (1953) originally suggested that mist pick-up may operate in the dispersal of the conidia of *Cercosporella herpotrichoides*. Later Davies (1959) demonstrated that a vigorous spray of water droplets could detach conidia in certain moulds. More recently mist pick-up has been the subject of careful experimental study by Dowding (1969). He used mists with droplets of defined size (2–60 μm) moving at velocities of from 0·48 to 10·3 m/s over colonies of the blue-stain fungus *Ceratocystis piceae*. After passage over the fungus the droplets of mist were impacted on malt agar in Petri dishes in an Andersen sampler. Colonies of *C. piceae*, derived apparently from the slimy conidial stage, developed on these plates. Even at the lowest velocity used (0·48 m/s) spores were picked up by the mist, but there was no liberation into dry air-streams. Similar experiments with dry-spore moulds (e.g. *Penicillium* sp. and *Cladosporium herbarum*) indicated that mist was no more efficient in dislodging spores than was dry air. In two species of *Chaetomium* it has been shown (Harvey, Hodgkiss, and Lewis 1969) that ascospores exuded through the ostiole of the perithecium (see p. 50) are not liberated into rapidly flowing air-streams (5 m/s) whether the air is dry or moist, but are set free in substantial numbers if mist droplets from an atomizer are injected into the air stream.

Most dry-spore moulds grow not on the ground, but on living plants or on dead vegetation, and any vigorous shaking produced by gusts of wind, heavy rain, or passing animals may be effective in spore liberation.

The diseased condition in man known as farmer's lung may be caused by inhaling fungal spores that are liberated in vast numbers during the handling of damp hay and straw. Clearly, therefore, the characteristics of spore liberation from these substrata are important in understanding the disease with a view to its control. Gregory and Lacey (1963) have studied this in some detail. Their method involved gentle shaking of the damp hay by revolving a sample (20–30 g) in a perforated drum in a wind tunnel through which air could be sucked at a controlled speed, spores being trapped downwind by a cascade impactor. The spores caught included those of *Cladosporium*, *Aspergillus*, *Penicillium*, and *Trichothecium*. The results of this work are in close agreement with those of Zoberi (1961) on liberation from dry-spore moulds not subjected to mechanical agitation. Spore release increased with the velocity of the air-stream in the wind tunnel. Further, most spores were liberated near the start of a test, the number set free falling off sharply with time.

10. SPORES IN THE AIR

GROWTH of moulds on damp bread, moist leather, and decaying vegetable matter must have been familiar to man from early times, but the recognition of these growths as fungi came only with the use of the compound microscope. Robert Hooke, the first of the great microscopists, suggested in 1665 that moulds are 'nothing else but several kinds of small and variously figur'd Mushroms', but even when recognized as fungi they were still generally regarded merely as by-products of organic decay without the power of reproduction. Clear evidence that these organisms develop from 'seeds' was given in the early eighteenth century by Micheli, who not only observed the microscopic 'seeds' but also showed by experiment that they could reproduce the species. The 'seeds' were, of course, spores, but the academic distinction between seeds and spores was made somewhat later by Hedwig. Towards the end of the eighteenth century Spallanzani, arguing from well-conceived controlled experiments, threw additional doubt on the generally accepted view of spontaneous generation and pointed out that the apparently spontaneous development of moulds was probably due to the omnipresence of their spores. He wrote: 'The seeds of mould may be disseminated in such abundance as to enter into the composition of all animal and vegetable substance . . . it is the readiest method of accounting for the extraordinary abundance and universal existence of mould.' Some eighty years later Pasteur proved, in so far as a negative can be proved, that neither fungi nor bacteria are generated spontaneously, but that their reproductive units normally occur in air, so that following brief exposure any suitable medium, although initially sterilized, may develop colonies of bacteria, yeasts, and moulds. This early history of aerobiology is fully discussed by Ramsbottom (1941) and by Gregory (1961).

The spore content of the air has been extensively studied. It is a matter of concern to the mycologist studing the epidemiology of plant diseases, and, to the medical man concerned with inhalent allergy in which pollen grains and fungal spores are of prime importance. The whole field of study is the subject of an outstanding book by Gregory (1961). He introduced the useful concept of the 'air spora' for 'the population of airborne particles of plant or animal origin'.

In the latter half of last century a number of workers (e.g. Pasteur (1862), Cunningham (c. 1873), and Miquel (1883)) made considerable contributions to this study of aerobiology. Then interest slackened until it was revived about the middle of the present century by plant pathologists (e.g. Stakman and Christensen (1946) and Gregory (1945)) and by the allergists (such as Durham (1942) and Hyde and Williams (e.g. 1946)).

One method of studying the air spora has been to trap the spores on a horizontal sticky slide, protected from rain but otherwise so placed that air has free access. After a period of exposure a defined area of the slide is scanned and the spores identified and counted. The only merit of this method is its simplicity; but for quantitative work it is of little value. The number of spores deposited on the slide depends on several variable factors. Under relatively still conditions the larger spores will sediment from the air more rapidly than the smaller, and turbulent conditions will also favour the impaction of the larger spores. Further, the volume of air sampled in a given time will depend on the velocity of the wind.

The use of a vertical slide, kept facing the wind by a vane, is preferable, but again impaction efficiency is low and varies greatly with spore size, and the catch cannot be related to a definite volume of air sampled, unless wind-speed is determined throughout the exposure.

There are two fundamental problems in studying the air spora on the basis of spores trapped on sticky slides. First there is the problem of identification since so many spores look so alike. However, with careful study and experience it is surprising how many can be identified to the genus if not to the species. Secondly, it is normally impossible to distinguish, from mere inspection, between living and dead spores. These limitations have led many to prefer to trap spores on surfaces of nutrient agar briefly exposed to the air in Petri dishes. Subsequent incubation then allows the spores to give rise to colonies which, if necessary after subculturing, can be identified. But here again there are difficulties. The nature of the nutrient medium differentially affects the species that develop; rapid growers tend to swamp those that grow more slowly; and many viable spores, notably those of obligate plant parasites, are unable to produce colonies on agar.

In spite of their shortcomings both these simple methods of sampling the air spora have given valuable information, particularly in relation to the broad trends in periodicity throughout the year and in studying the occurrence of spores in the upper air sampled from aircraft. Nevertheless, with the development of satisfactory volumetric traps the tendency now is to concentrate on more modern methods in which the number of spores in unit volume (usually a cubic metre) of air can be reliably estimated. It is now necessary to review the principle techniques of such volumetric sampling.

One of the earliest of the volumetric traps was Pasteur's gun-cotton filter. This was elaborated by Frankland (1887) for the quantitative estimation of micro-organisms in the atmosphere. A known volume of air was aspirated through a tube stuffed with gun-cotton which filtered out suspended particles. This was later dissolved in an ether–alcohol mixture and the spores in the liquid were examined.

A modern apparatus designed to deal with the same problem is the 'cascade impactor' developed by May (1945) in connection with the analysis of particle size in fogs and aerial dusts (Fig. 10.1). Air, sucked

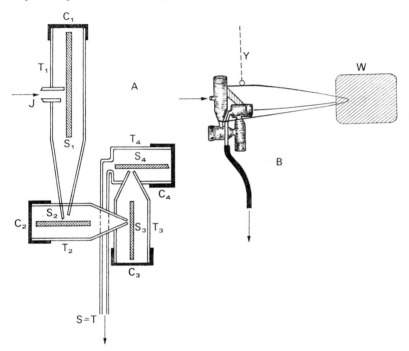

FIG. 10.1. Cascade impactor. (A) diagram of apparatus seen in longitudinal section; J, entry jet; S_1–S_4, greased sampling slides; T_1–T_4, metal tubes fitted with movable caps (C_1–C_4); S–T, suction tube; (B) whole apparatus on smaller scale, showing the wind-vane attachment (W) and the thread (Y) by which the apparatus is suspended. Arrows show movement of air-stream sucked through the apparatus. Based on May (1945).

through the whole apparatus at a controllable speed, impinges as jets successively on four sticky slides; the air, as it were, cascades from one slide to the next. The largest particles are deposited on the first slide, smaller ones on the second, and so on. The range of size of particle impacted on each slide depends on the size and density of the particles and on the rate of air-flow through the system. With an intake of 17·5 l/min and with particles having a density of 1·0, the first slide traps those of 5·5–20 μm diam; the second particles 1·5–7 μm diam; the third those of 1–3 μm diam, and the fourth still smaller ones. Clearly fungal spores should be deposited on the first two slides. At the end of a run, the slides are removed and the spores deposited on each are counted. Knowing the rate of flow of the air through the trap and the period of operation, the counts can be translated into numbers per unit volume of air.

There are certain important features in the construction and operation of such an impactor. First, the form of the intake orifice is important; it must have thin walls with sharp leading edges. Secondly, the intake must directly face into the wind. This is achieved by the use of a vane. Thirdly, the actual flow through the intake orifice, produced by the suction, must be at a rate roughly equal to the outside wind velocity, otherwise the entering streamlines are distorted, leading to significant errors. In operating a trap of this kind suction is usually adjusted to give an inflow at a speed corresponding to the average wind velocity.

As a development from the cascade impactor, the Hirst spore trap has now become the standard apparatus used in most aerobiological studies (Hirst 1952). This is essentially like a single stage (actually the second) of the cascade impactor, but the sticky slide on which the entering jet of air impinges is moved by a clockwork mechanism past the orifice at 2 mm/h. On removal of the slide from the trap it can be scanned under the high power of the microscope and by studying traverses 4 mm apart (corresponding to 2 h) changes in the air spora during the course of the day can be recorded. The essential principles, though not the detailed construction, of the Hirst trap are illustrated in Fig. 10.2.

Volumetric suction traps depending on impaction, but involving Petri dishes containing nutrient agar instead of greased slides, have also been developed. They are of value in certain investigations when interest centres on rapidly growing and freely sporulating saprophytic fungi (especially moulds) and when it is important to know that the trapped spores are viable. This type of spore trap was developed by Bourdillon and his co-workers (1941) and other investigators have used essentially the same apparatus in a modified form. It is illustrated in Fig. 10.3. Air is drawn at a measured speed and for a given time through the system and thus the volume concerned can be calculated. The air enters by a narrow slit, at the end of a movable metal tube, adjusted to be close to the surface of sterile nutrient agar in a Petri dish placed on a turn-table which is slowly rotated during the sampling period. The dish is then removed, covered, and incubated. Subsequently the developing colonies are identified and counted. From the data obtained the concentration of viable spores of each type in the air can be deduced.

In the Andersen sampler (Andersen 1958) the 'cascade' principle is used with the spores impacted on agar in Petri dishes. Six are arranged one above the other and separated by metal disks each perforated by several hundred small holes of decreasing size from top to bottom of the stack. The whole forms a closed cylinder through which air is drawn at

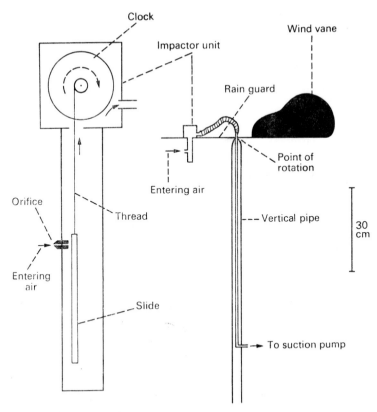

FIG. 10.2. Hirst spore-trap. *Right:* set-up seen in vertical section. *Left:* impactor unit at a larger scale. Arrows indicate how the air circulates. Very diagrammatic.

about 30 l/min (Fig. 10.4). Having passed through the first perforated disk, the air spreads out over the agar in the first dish, depositing the largest spores, and spills over its rim before passing through the next disk and over the second agar surface; and so on, progressively smaller spores being impacted on successive plates. The dishes are then removed, covered, and incubated, and the developing colonies are counted and identified if possible. From the data obtained the number of viable spores in unit volume of air can be calculated.

The rotorod (Fig. 10.5) is a volumetric sampling device of a rather different nature. It consists of a U-shaped structure (with arms 6 cm high) usually made of brass wire of square cross-section (1·6 × 1·6 mm). This is attached upright at its mid point to the vertical axis of a small electric motor (run off a dry battery) and capable of rotating at 2400 rev/min. The nearly vertical arms of the U can either be replaced with clear plastic of the same dimensions as the wire and made sticky with vaseline,

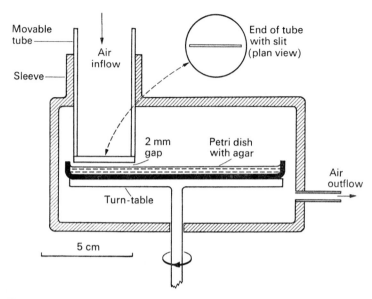

FIG. 10.3. Bourdillon slit-sampler. Diagram of the essential features. Slit 0·25 mm wide, 3·0 mm deep, 275 mm long. Turn-table rotated slowly during sampling period. Slit 2 mm above surface of agar. Access to sampling chamber through lateral window. Exposure time several minutes. Based on Bourdillon *et al.* 1941.

or the leading surface of both arms can be furnished with a narrow vase-lined strip of transparent cellophane or glass suitably attached. During rotation, spores are impacted on the sticky surface. Later each plastic rod or vaselined strip can be mounted horizontally on a slide and the spores on an appropriate area can be identified and counted. Knowing the area of surface under consideration and the speed at which rotation occurred, the volume of air sampled can be calculated.

In the rotorod there is remarkably efficient impaction of spores and, provided the air is itself in reasonable motion, reliable results are obtained. It samples up to 100 l/min. A great merit is its simplicity and portability. It can be assembled in an open-work plastic basket (Gregory and Lacey 1964) and can be carried through the woods during a foray in the autumn, when the mycologist ends up with a collection of fungal spores instead of a basketful of toadstools.

Although most sampling of the air spora has been at or near ground level, there has been considerable study of the microbial content of the air at much greater heights in the trophosphere, which extends up to about 10 km. Most modern work has involved trapping spores from aeroplanes. Many kinds of collecting device have been used including sticky slides and Petri dishes with agar, but it is only recently that suction impactor

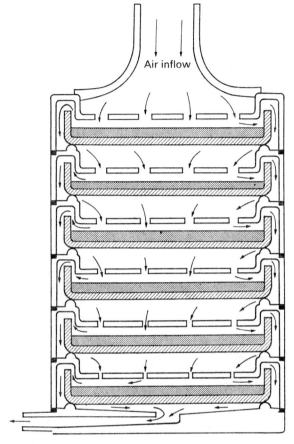

Air inflow

FIG. 10.4. Andersen sampler. Simplified diagram of sectional view of assembled sampler. Each metal segment (involving a perforated disk) separated from the next with a rubber ring (shown black). Each half Petri dish (shaded) contains agar (stippled). Flow of air through sampler indicated by arrows. Certain constructional details omitted.

traps have been employed with flow through the intake orifice rendered isokinetic by equating its velocity to the speed of the aircraft. Such traps allow concentrations of spores to be estimated as numbers per m^3 of the air (Hirst, Stedman, and Hogg 1967).

In the stratosphere (above 10 km) knowledge of the spora has come from very sophisticated samplers carried as high as 25 km by balloons. Since spores seem to become increasingly rare with height, enormous volumes of air must be filtered through the apparatus and precautions against the true picture being distorted by stray contamination, on the ground and during passage through the trophosphere, have to be extremely rigorous.

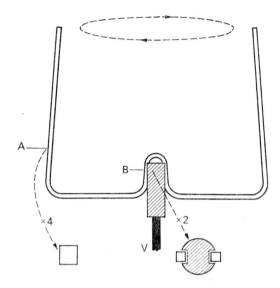

Fig. 10.5. Rotorod sampler. Semi-vertical collector arms 6 cm high, 0.16 cm diam. Vertical axis (V) attached to small electric motor (not shown). Cross-section at (A) enlarged to four times true scale and at (B) to twice the scale of the rest. Speed of motor 2400 rev/min.

Now bioprobes are being developed for use from rockets and space vehicles operating at much greater altitudes and in space itself (Bruch 1967).

Having discussed the methodology of aerobiology, it is desirable to consider in broad outline the nature of the air spora from the mycological point of view (Fig. 10.6). In many parts of the world the commonest spore-type belongs to *Cladosporium*, and most records seem to be of *C. herbarum*, sometimes present to the extent of several thousand spores per cubic metre of air. Again ballistospores of the mirror-image yeasts (especially *Sporobolomyces*) are very abundant. Other common spores in outdoor air are the conidia of *Alternaria*, *Botrytis*, and *Epicoccum* and, in the summer, urediospores of rusts, chlamydospores of smuts, conidia of powdery mildews (Erysiphales), and sporangia of downy mildews (Peronsporales). Conidia of *Penicillium* and *Aspergillus* and spores of Mucorales make little contribution to the spora of the air out-of-doors, although they may be significant in indoor air.

Early studies on spores in the air, using the exposed Petri dish method, failed to reveal the presence of Basidiomycetes, which constitute the most conspicuous elements of our fungal flora. This was probably because basidiospores tend to show a lag in germination and because the colonies grow slowly. However, examination of slides exposed in a Hirst trap often reveal many basidiospores, apart from the ballistospores of *Sporobolomyces*

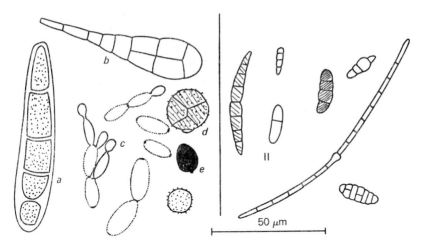

FIG. 10.6. *Left*, common types of fungal spore trapped on rotorod in garden in Kent, England in May 1969 in early afternoon on a fine day. (*a*) *Helminthosporium;* (*b*) *Alternaria;* (*c*) *Cladosporium* (single spores and groups of spores, commonest type in the spora); (*d*) *Epicoccum* (*e*) basidiospore of *Coprinus;* (*f*) spore of *Reticularia* (a slime mould). *Right*, common types of septate ascospore trapped in early afternoon of next day following onset of rain; *Cladosporium* also present in small amount but little else of normal dry-air spora.

These may, indeed, be present in sufficient concentrations to be important medically in connection with inhalent allergy. Thus the school of Hyde and Williams, which has done so much valuable work on this subject in South Wales, has shown that great numbers of basidiospores occur in the air in summertime, the numbers rising as the early autumn agaric climax approaches (Fig. 10.7). Naturally the concentrations are much higher in the air of a wood as compared with that of a city (Adams, Hyde, and Williams 1968).

Not only does the spora, particularly in temperate regions, vary in composition throughout the year, but it goes through a daily cycle mainly associated with differences in timing of take-off in various fungi. This circadian periodicity will be discussed in a later chapter. Further, a dry air-spora and a damp air-spora can be recognized (Hirst 1953). Following rain there tends to be a large element of elongated ascospores discharged from perithecia shortly after wetting, and rain also tends to comb the existing spores out of the air, bringing them to earth. The consequence is that the spore catch from damp air on an impactor slide may present a very different picture from that seen when the air is sampled in dry weather (Fig. 10.6).

Normally the spores caught on sticky slides in spore traps occur singly and without directly associated debris, suggesting that most have become incorporated in the air by direct take-off and not indirectly by, for example, being derived from soil dust. Indeed, it would seem that quite a high proportion may become airborne by violent discharge.

In recent years there has been considerable interest in the dispersal of airborne spores around a centre of liberation, for this is no mere academic question. It is, for example, important to a plant pathologist attempting to define the practical limits of a danger zone round a locus of infection.

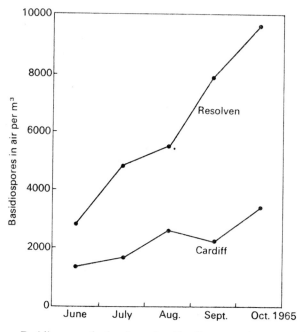

FIG. 10.7. Basidiospores, both coloured and hyaline, 13 m above ground level in a park in central Cardiff, and at 2 m above ground on edge of a spruce plantation at Resolven about 40 km WNW of Cardiff, Wales. Value for each month is an average of the daily count per m³ air. Hirst spore-trap used. Based on data from Adams *et al.* (1968).

Indeed, the general form of airborne infection around a source is well known (Fig. 10.8). The curve is a hollow one with infection falling off very steeply as distance increases.

The modern view pictures dispersal from a point source steadily liberating airborne spores as rather like the plume of smoke from a tall factory chimney. Under average conditions such a plume takes the form of a cone of small angle with its apex at the mouth of the chimney. As the cloud moves down wind it is diluted and widened by mixing with peripheral eddies of smoke-free air. The mean concentration of particles at any cross-section is inversely proportioned to the distance from the

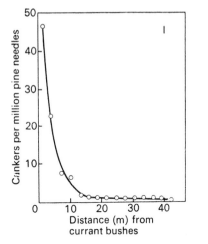
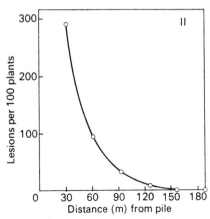

FIG. 10.8. (I) *Cronartium ribicola.* Cankers formed on pine needles near heavily infected currant bushes. Data from Buchanan and Kimmey (1938). (II) *Phytophthora infestans.* Blight in potato field near infected potato-refuse pile. Data from Bonde and Schultz (1943).

source, but concentrations will tend to be greater near the horizontal axis of the cone than towards its surface. Under more unstable conditions, as on a sunny afternoon in summer, the smoke plume is strongly looped and, as distance increases, discrete clouds of smoke may be formed (Fig. 10.9). Possibly the conical model of dispersal of smoke or spores may be valid only for relatively short distances from the source.

In nature a spore cloud is usually generated from a source near ground level and in consequence the base of the horizontal plume, much too dilute to be visible as such, drags along the surface, losing spores steadily by various types of deposition as it proceeds.

To Stepanov (1935) we owe pioneer experimental studies on deposition round a centre of spore liberation. A more recent study by Sreeramulu and Ramalingam (1961) may be considered in some detail. Their experiments were essentially of the Stepanov type. They liberated a mixture consisting of 9.39×10^8 *Lycopodium* spores (32.3 μm diam) and 9×10^9 *Podaxis* spores (14×11 μm diam) at a height of 0.5 m. Spores were trapped on vaselined horizontal slides placed on the ground along a series of radii extending down wind for about 35 m and 30–50° on either side of this direction. The results of one of their experiments are summarized in Fig. 10.10. Neglecting the value for deposition at 2.5 m, the hollow curve, which has already been noted in connection with airborne pathogens, is clearly shown. The lower value at 2.5 m than at 5 m can be related to the source being 0.5 m above ground, and this no doubt also has a significant effect in lowering deposition at 5 m.

We shall return later to the difference in rate of deposition of the larger *Lycopodium* spores and the smaller ones of *Podaxis*, but now the theory of dispersal around a centre of spore liberation requires brief mention.

Since wind is so capricious in its behaviour, it may well be doubted if any quantitative generalizations can be made, but Gregory (1945) has pointed out that 'the problem is statistical because, whilst the destination

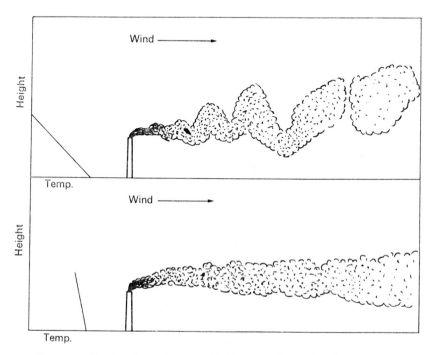

Fig. 10.9. Smoke plume from a tall chimney. *Above:* (marked decrease of temperature with height) plume strongly looped and soon becomes broken up. *Below:* (only slight decrease of temperature with height) plume conical. Based on Geiger (1965) after Wexler.

of a single spore in a wind eddy is inscrutable, the average distribution of a vast number of spores over a uniform area and over a period of time offers hope of rational treatment.' He has applied modern meteorological concepts of eddy diffusion to the problem and has developed a somewhat elaborate formula to describe dispersion around a point source. The great advance of this over earlier attempts at mathematical formulation of the problem is that the constants used are not arbitrary, but have a defined physical significance. Gregory claims a reasonable agreement between theory and fact. Schrödter (1960) has also considered the mathematics of dispersal in considerable detail. Gregory's formula takes no account of spore size, since it is argued that the rate of fall of

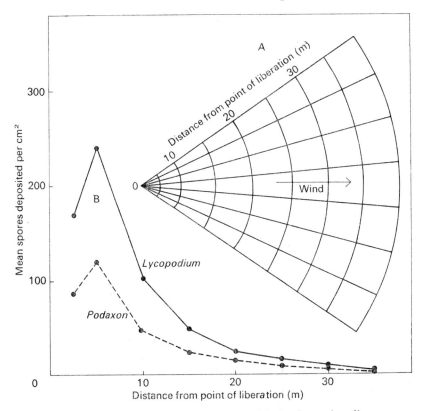

FIG. 10 10. Spore deposition on horizontal slides at increasing distances from point of liberation 0·5 m above ground. (A) layout of experiment; slides exposed at all points of intersection of radii and arcs shown. (B) spore deposition of *Lycopodium* and *Podaxis* spores at increasing distances (mean values for indicated arcs).

spores is insignificant in relation to the movements that occur within the eddies of an air mass in turbulent motion. The rate of fall in still air varies from about 0·05 cm/s for the minute spores of *Lycoperdon pyriforme* to 2·5 cm/s for the very large ones of *Helminthosporium sativum*. It is worth noting that this latter terminal velocity is small compared with that of the caryopsis of thistle (*Cirsium arvense*) with a rate of fall in still air of 16·8 cm/s.

Schrödter (1960) criticizes the neglect of rate of spore fall as a significant factor in aerial dispersion. He says 'The velocity of fall is an extremely important factor in determining the range of flight and cannot be neglected in the problem of dissemination.' It is obvious that, in the dispersal of propagules liberated into the air by plants, size must become important at some level. The difference between the potential for dispersal of an acorn and a pine seed is obvious. The question is whether there is any significance in differences in size at the level of microscopic particles.

In Sreeramulu and Ramalingam's experiment, in spite of the fact that the ratio of the number of the small spores of *Podaxis* liberated to that of the big ones of *Lycopodium* was ten to one, deposition of *Lycopodium* spores on the slides down wind was more than twice as great as the deposition of *Podaxis* spores. They concluded that 'the rate of deposition of the two types from suspension in the spore cloud was directly proportional to the individual spore volumes.' These results demonstrated that spore size is certainly an important factor in deposition if not in actual dispersal, but it is to be remembered that in deposition both sedimentation under gravity and impaction under the influence of eddies are involved, and in both processes the larger spores are favoured. Clear evidence of spore size being significant in dispersal itself comes from the work of Hirst *et al.* which will be discussed shortly.

In considering decrease of spore concentration on receding from a centre of liberation, attention has been fixed on horizontal gradients, but in addition there are vertical gradients. The earlier workers, starting with the pioneer work of Stakman and his colleagues (1923), collected spores on sticky slides exposed from aeroplanes. They invariably noted a striking decrease in catch with height. For example, Craigie (1945) examined the air above wheat fields in Manitoba heavily infected with stem rust. Each slide was exposed from the aircraft for 10 min and the number of urediospores caught per square inch was determined (Table 10.1). The rapid

TABLE 10.1

Date	Number of urediospores per in^2 intercepted by microscope slides at different heights: 10-min exposure			
	1000 ft	5000 ft	10 000 ft	14 000 ft
5 August 1930	24 200	7560	108	10
10 August 1930	24 000	170	36	30

decrease of spore concentration with increasing altitude is evident and is probably a substantially true picture in spite of the unsatisfactory method of sampling. Now, with the development of volumetric traps adapted for use from aircraft, these gradients have been more fully analysed and the general decrease of spore concentration with height has been confirmed. This decrease is normally logarithmic, although the picture varies in different conditions of air stability (Fig. 10.11).

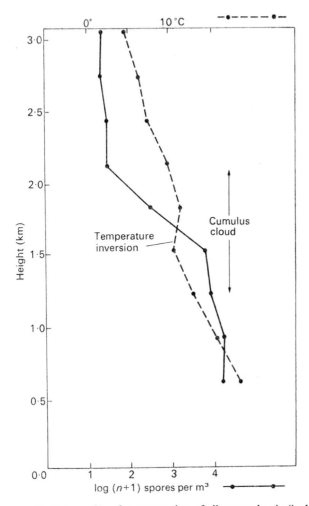

FIG. 10.11. Verticle profile of concentration of all spores in air (including pollen) under stable air conditions sampled by aircraft in mid-afternoon (14.30) in August over the land. Temperature profile indicated. After Hirst *et al.* (1967*a*).

Hirst and his colleagues (1967) have studied the spora of the air above the North Sea in relation both to distance from the land and to height. As relatively spore-free air masses from the Atlantic drift eastwards over Britain they recruit spores and pollens mainly by day through the agency of upward currents due to thermal turbulence. Thereafter the air masses bearing their British spores continue their eastward passage over the sea. In the space of a few hours the vertical and horizontal spore profiles at varying heights up to 2000 m and along a line stretching

ENE from the Yorkshire coast to the Skagerrak were sampled from an aircraft. For particular days profiles have been prepared showing the distribution of equal concentrations of spores (isospores) related to distance from the English coast and to height above the sea.

Figure 10.12 gives the picture for *Cladosporium*. On the summer day in question there was a 'cloud' of spores, with concentrations above 160

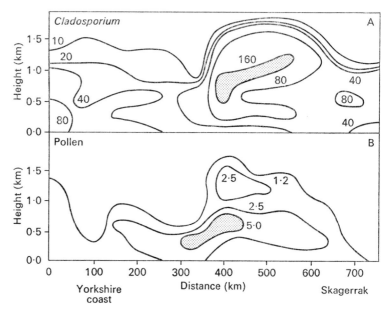

FIG. 10.12. Distribution of *Cladosporium* spores and pollen grains on 16 July 1964 in relation to height above the sea, and distance between the Yorkshire Coast and the Skagerrak north of Denmark. Isospore lines (indicating number of spores per cubic metre) are separated on a geometric progression increasing by a factor of two. Region of highest concentration stippled to indicate the major 'spore clouds'. Vertical scale greatly exaggerated in relation to the horizontal one; both in kilometres. Slightly modified after Hirst *et al.* (1967).

spores/m^3, at a height of around 1000 m. Evidence of the rate of movement of the air suggested that the spores in this cloud became airborne about a day earlier. The existence of such a more or less discrete cloud reflects the diurnal periodicity of incorporation of spores in the air over land. It is of interest that at the same time the corresponding spore cloud of the much larger pollen grains was at a lower level, about 500 m, suggesting the importance of spore size in relation to dispersal, at least on this particular day.

The concept of spores moving with definite air masses has been particularly emphasized by Pady and his co-workers. Thus Pady and Kapica

(1955) examined the spores over the Atlantic Ocean at a height of 2500–3000 m. The catch obtained seemed to depend on the origin of the air masses sampled, much higher concentrations being found in tropical air that had previously drifted over land masses bearing rich vegetation, than in polar air that had passed over less land and that relatively unproductive of spores.

Clearly the transport of spores in the air is largely concerned with the movement of air masses, not only on the grand scale, as with tropical and polar air over the Atlantic, but also on a smaller scale. The local spread of the white-pine blister rust (*Cronartium ribicola*) from *Ribes* to *Pinus strobus* in the Great Lakes region of the United States is a case in point (van Arsdel 1965). The pines just above the *Ribes* bushes growing in the lower swampy regions are relatively free from infection; it is those higher up that are attacked. Basidiospores are released at night, with a peak rate around 1 a.m., into an air-stream flowing down the slope. This stream then rises, in forest patches in the swamp at the base of the slope, and becomes involved in a return flow above a temperature inversion thus bringing the basidiospores to the pines at a greater altitude (Fig. 10.12). This behaviour of the air, which accounts for the infection pattern, has been studied experimentally by observing the movement of smoke released from grenades.

The possible presence of micro-organisms at very high levels in the atmosphere has assumed considerable importance in connection with space programmes. It is becoming increasingly necessary to determine the vertical extent of the 'biosphere' where viable spores and micro-organisms occur. It is important to know at what height a space craft can safely reject its protective canister so that it can continue its flight in a sterile condition and not pick up organisms of terrestrial origin and introduce these to the moon or to a planet.

Since aeroplanes cannot operate much about 40 000 ft, exploration at greater heights depends on balloons (up to about 90 000 ft), or on rockets or space craft launched by them. Investigations of this nature are extremely exacting, involving the problem of sampling of very considerable volumes of the atmosphere or of space and avoiding completely all sources of stray contamination.

From balloons viable spores have been caught at 60 000–90 000 ft by spore traps but the concentration is extremely low; about one spore per 2000 ft^3 sampled. *Cladosporium* and *Alternaria* have been consistently recovered from these heights. Now volumetric impactor traps have been devised for rockets and space craft and information should soon be available concerning the presence or absence of spores at much greater altitudes (Bruch 1967).

The possibility of the presence of micro-organisms in space beyond the earth's atmosphere is now taken quite seriously (Sussman and Halvorson

1966). Long ago Arrhenius suggested the concept of *panspermia*, the migration of spores from one planet to another under such influences as light pressure. Some are now giving this long-discarded concept a new lease of life, and one basic objection that spores could not possibly survive the rigours of space is now no longer so firmly based.

Returning to earth, it is to be emphasized that the principal fact emerging from a study of wind dispersal from a centre is that it is essentially local, and the plant pathologist normally finds that the practical

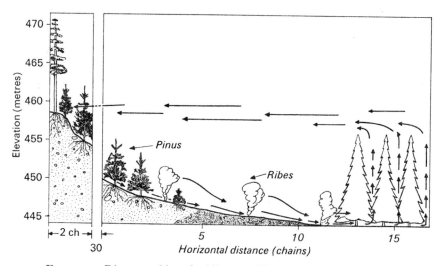

Fig. 10.13. Diagram of how basidiospores liberated by night from *Ribes* bushes (near the margins of a swamp) are carried to *Pinus* on the higher slopes by the local air-circulation pattern established at night-time. After van Arsdel (1965).

safety zones round centres of infection are remarkably small. However, the spread of cereal rusts calls for special attention, since in certain regions long-distance infection on a considerable scale may frequently occur.

The epidemiology of black stem rust of wheat caused by *Puccinia graminis tritici* has been extensively studied, especially in North America where there is a fairly continuous belt of wheat cultivation spreading northwards from Mexico, through the Mississippi valley, to western Canada. There infection of the wheat in early summer may be by aecio-spores from local barberry bushes, by urediospores that may have survived the winter, or by airborne urediospores coming from far away.

In connection with the epidemiology of black stem rust Craigie (1945) made a long-term study of urediospores in the air at a number of stations in western Canada. The technique used was to expose sticky glass slides each mounted on a vane so that they continually faced the wind.

and to record the spores deposited on these. It was found that viable spores occurred in the air of Manitoba for a week or two each year *before* any infections could be seen in the field. Sometimes the slides showed sudden, but temporary, increases in urediospore deposition associated with 'spore showers'. A shower was usually followed, after a necessary incubation period of a week or two, by the development of rust on the wheat in the neighbourhood. Records of an unusually early and spectacular shower in mid-June 1929 are given in Table 10.2. It occurred between 16 and 18 June. The evidence suggested that the urediospores came from a region of

TABLE 10.2

Date	Number of stem rust urediospores caught each 2-day period on 1 in^2 of sticky glass surface at		
	Morden	Winnipeg	Brandon
12 June	0	0	0
14 June	0	0	0
16 June	24	133	0
18 June	174	193	68
20 June	0	0	0
22 June	0	0	0
24 June	0	0	0
26 June	0	0	0
28 June	15	1	0

the Mississippi valley at least 500 miles away, since at that time the northern limit of rusted wheat was approximately along the line shown in Fig. 10.14. A number of similar examples of long-distance dispersal of rust spores are also well documented, not only from America but from other parts of the world including Europe (Zadoks 1965), and this raises the question of whether it is different in kind from local dispersal round a centre. It is difficult to answer this, but it should be noted that the urediospores certainly do not come from a 'point' source, but perhaps from an area of rusted wheat of several thousands of square miles. Again the picture of the steady dilution of a spore cloud with distance inherent in the conical model of dispersal may not be entirely valid, and spore clouds, as recognized by Hirst and his co-workers, may retain a certain identity and travel considerable distances before depositing their load.

The fact of long-distance dispersal of the spores of wheat rust raises the issue of the possible spread of viable spores across the Atlantic between America and Europe. The available evidence still suggests that the Atlantic acts as an effective barrier to the spread of airborne fungal diseases of plants.

So far dispersal of spores through the air has been discussed without reference to their viability, but the fact that a spore may be transported a long distance is of no significance in the biology of the fungus if at the end of its journey it is incapable of germination.

The principal factors that may lead to loss of viability during passage through the air are dessication and the injurious effects of radiation particularly ultraviolet light. Thin-walled spores seem especially susceptible,

FIG. 10.14. Sketch map of N. America showing northern limit of rusted wheat in mid-June 1929, when a very early shower of urediospores fell in western Canada at Brandon, Morden, and Winnipeg.

while those with thicker walls are more resistant to drying, although it is difficult to define why this should be so. It is more understandable that spores with pigmented walls may be more resistant to the effects of light than are those with transparent walls. However, detailed evidence on this matter is scarce.

Workers on the spread of diseases caused by rust fungi generally agree that the basidiospores (having thin transparent walls) are normally capable only of short-range infection, while urediospores (having relatively thick, pigmented walls) can, as we have just seen, spread disease over much greater distances in a single hop. This difference is often attributed to differences in the viability of the spores during transport through the air.

A number of workers have given special attention to the problem of the viability of the elements of the air spora. Viability can, in general, be determined only by the ability to demonstrate germination, but failure to induce a spore to germinate, under the particular experimental conditions used, does not necessarily mean that it is not viable.

Viability of the air spora has been studied by Pathak and Pady (1965) and Kramer and Pady (1968) working in Kansas, and by Davies (1957) in London. *Alternaria* spores especially appear to withstand the rigours of aerial dispersal with an average percentage germination of 80 per cent. The Kansas workers found less viability in *Cladosporium*, with means of 33 to 52 per cent in different tests, but Davies obtained much higher values, often over 90 per cent, in the city air of London. The difference may not, however, be real, but may reflect the use of different methods for determining the ability to germinate.

Pathak and Pady have found that viability may vary with the time of day when spores are trapped. A striking example (Fig. 10.15) relates to fusiform ascospores with a high viability of 60–80 per cent in the early hours of morning falling to roughly half that value in the afternoon. Spores are much more numerous in the air around dawn, perhaps because spore release tends to be nocturnal. The decrease in apparent viability after noon may be due to the fact that the spores trapped at that time of day have been longer suspended in the air.

In the general story of a fungus dispersed through the air there are, as with an aircraft, three major episodes: the take-off, the flight, and the landing. Much of this book is concerned with the first of these because this has been the author's particular interest; the present chapter has been, so far, largely concerned with the flight, but now something must be said about the touch-down.

The coming to rest of spores on a surface is referred to as deposition and, under dry conditions, this may occur by sedimentation under the influence of gravity or by impaction when they are deposited by the action of air currents. However, rarely do these two processes act alone. Again, spores may be be brought to earth by falling rain.

Spore deposition has been the subject of considerable investigation. Using spore clouds of controlled concentration Gregory (1951, 1961) has studied the impaction of spores on sticky cylinders and microscope slides in a wind tunnel. In considering impaction on an object a measure of efficiency is important. Gregory defines this (E) as a percentage,

$$E = \frac{\text{trap dose (T.D.)}}{\text{area dose (A.D.)}} \times 100.$$

The 'area dose' refers to the number of spores that would flow through 1 cm^2 at right angles to the wind in the trapping zone (if the trap were not present) during the course of the experiment, and the 'trap dose' is

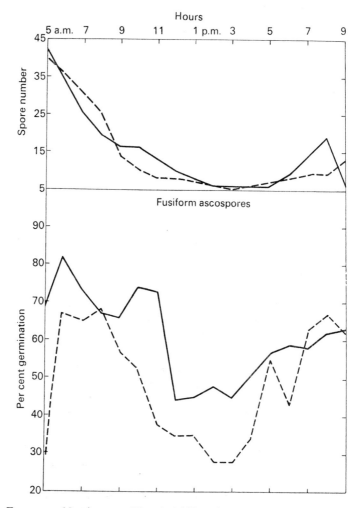

FIG. 10.15. Numbers per ft³ and viability of airborne fusiform ascospores (thin-walled, hyaline to sub-hyaline). Two series of samples, solid line mean of 20 samples taken in summer 1959, dotted line mean of 18 samples for summer 1960. After Pathak and Pady (1965).

the spore deposition per cm² of trapping surface over which the spore cloud passes.

Using cylinders, Gregory found the efficiency of trapping of *Lycopodium* spores increased with wind speed and decreased with diameter (Fig. 10.16). Considering spores of different size, larger ones were more readily impacted than smaller ones (Gregory 1951). Using the minute spores of *Lycoperdon giganteum* (4·5 μm diam) no impaction occurred under the conditions used for *Lycopodium* spores (32 μm diam).

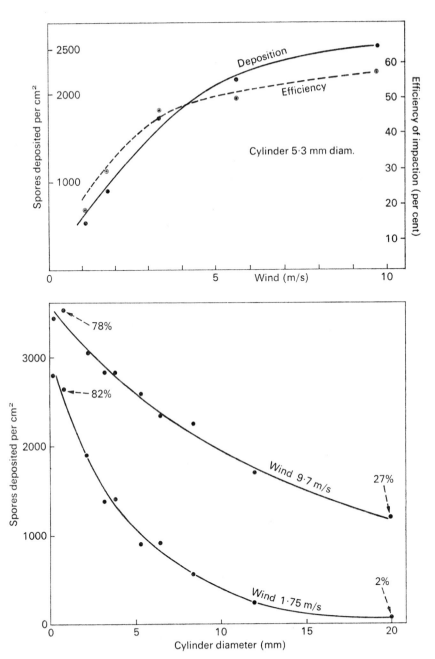

FIG. 10.16. Number of *Lycopodium* spores deposited 1.4 m from point of liberation on sticky cylinders per cm² (diameter × length) of presentation area per million spores liberated on axis of wind tunnel (with turbulent flow). To convert to deposition figure to mean density on upwind half-surface of cylinder, value must be multiplied by 0·5 π. *Above:* Deposition in relation to wind velocity on a cylinder 5·3 mm diam. The efficiency of impaction at different wind velocities also shown. *Below:* Deposition in relation to diameter of cylinder at two wind speeds. For 0·8 mm and 20·0 mm cylinders the efficiency of impaction at the two wind speeds is indicated. Based on data by Gregory 1961.

Some of the features of spore behaviour in a steady air-stream flowing past a cylinder are illustrated in Fig. 10.17. A small spore (3 μm diam), even when heading almost directly for the axis of the cylinder, may not be impacted (*a*), but be carried round the cylinder in the normal flow lines of the air-stream. However, a larger spore (10 μm diam) with a similar trajectory may have sufficient momentum to transgress the lines of flow

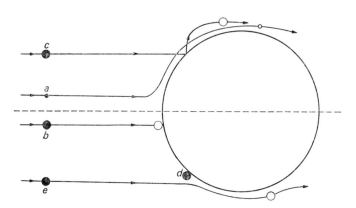

FIG. 10.17. Diagram of the trapping of dry spores on a dry cylinder (seen in transverse section) in an air-stream with laminar (non-turbulent) flow. Air contains large spores (10 μm) and small ones (3 μm). A black dot represents a spore at an earlier stage and a white one at a later stage. The small spore *a* is not impacted but the larger *b* on the same course is impacted. The large spore *c* bounces off and is carried round the cylinder but if it is already on the cylinder *d* it is not blown off. The large spore *e* is deflected round the cylinder without hitting it.

and be impacted (*b*). A large spore heading for the cylinder somewhat more tangentially (*c*) may still strike it, and remain there if the cylinder is sticky, but bounce off and be carried past if it is dry. However, if the spore is on a still more tangential course, it may never touch the cylinder (*e*).

Although the basic principles of impaction have been worked out in relation to cylinders, there is every reason to suppose that they are applicable in a general manner to objects of other form. Carter (1965) has studied deposition of ascospore octads of *Eutypa armeniacae* on leafy shoots of the host plant (apricot) in a wind tunnel. Although the efficiency of deposition was low, it varied greatly according to whether the spores came to rest on stem, petiole, or leaf lamina. It is to be noted that the diameter of the petiole was about 1 mm and that of the stem about 5 mm. As expected from Gregory's experiments with cylinders, the efficiency of impaction increased with wind speed, and petioles being narrower caught proportionately more spores than the thicker stems (Fig. 10.18).

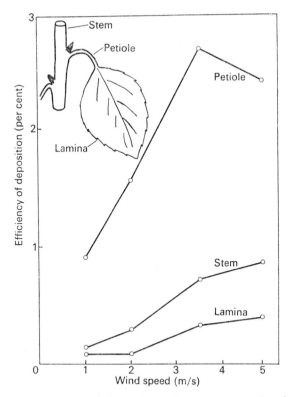

FIG. 10.18. Efficiency of impaction of ascospore octads of *Eutypa armeniacae* on stem, petiole, and leaf lamina of apricot (*Prunus armeniaca*). Based on graphs by Carter (1965).

A consideration of the spore-size factor in relation to impaction led Gregory to recognize certain spores as essentially 'impactors'. He pointed out that the spores of many stem and leaf parasites, such as the conidia of Erysiphales, the sporangia of downy mildews (Peronosporales), the urediospores of rusts, and the conidia of *Helminthosporium*, being relatively large, are good impactors and therefore capable of effecting a landing on their hosts. The spores of soil fungi, such as *Penicillium*, *Aspergillus*, *Mucor*, and *Trichoderma*, tend to be relatively small and must be brought to earth normally by processes other than impaction.

Under conditions of dead calm, and particularly at night in fine weather when the boundary layer of still air may be measured in metres rather than millimetres, spores may reach the ground or be deposited on vegetation by sedimentation under the influence of gravity. Then because of the greater rate of fall of larger as compared with smaller spores, sedimentation, like impaction, gives over-representation of larger spores when the air spora is sampled with a horizontal slide.

For ease of discussion impaction and sedimentation have been treated separately, but, as already mentioned, in most natural situations the two processes are operating simultaneously. Perhaps this can be illustrated best by considering spores trapped by a sticky horizontal glass slide held a short distance above the ground with both its surfaces exposed. Normally both will collect spores, but usually many more occur on the upper surface, where sedimentation and impaction operate, than on the lower where only the latter can occur. Under perfectly still conditions spores occur only on the upper surface; there is no impaction. In a highly turbulent situation equal numbers may be found above and below, all

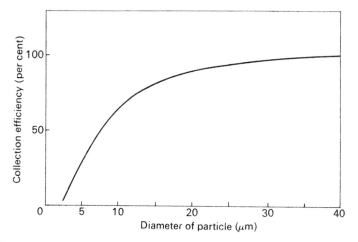

FIG. 10.19. Efficiency of collection of particles of increasing diameter by rain drops 2 mm in diameter. After Chamberlain (1967) p. 155.

spores having been deposited by impaction. Actually the picture of deposition on a horizontal slide, as indeed on any structure, is complicated by eddies developed in relation to the slide itself (Gregory and Stedman 1953).

Another mechanism by which spores may reach the ground is through the agency of raindrops. Now that there are available many records of the composition of the air spora throughout the day accompanied by parallel meteorological observations, it is clear that heavy rain often has a dramatic effect in sweeping spores out of the air so that immediately after a downpour the spora may be very thin (e.g. Sreeramulu and Ramalingam 1964, Pady *et al.* 1967).

The collection of spores suspended in the air by falling raindrops is essentially the same sort of process as the impaction of moving spores on stationary cylinders in Gregory's experiments. Large raindrops pick up big spores with reasonable efficiency, but smaller ones fail to be collected. Thus spores below 5 μm in diameter do not make contact with the falling raindrops. A curve based on the appropriate equation is shown in Fig. 10.19 and experiments give reasonable agreement with theory (Chamberlain 1967).

It has been the object of this chapter to review the general problem of aerial dispersal in fungi, a subject of special concern to the plant pathologist. The literature dealing with this matter is now voluminous, and reference has been made only to the more outstanding contributions or to those that have some special relevance for a particular aerobiological topic under immediate discussion. Finally, it must be emphasized that the student, in evaluating the results of any investigation, must make a critical appraisal of the methods used.

II. WATER-SUPPLY AND SPORE DISCHARGE IN TERRESTRIAL FUNGI

APART from the few fungi in which drastic drying of living cells operates the mechanism of spore release (pp. 128–137), turgid cells are essentially involved in discharge. This is true in Ascomycetes and Basidiomycetes and also in those fungi in which sudden rounding-off is responsible for the violent liberation of the spores. It is clear, therefore, that if continued spore release in these fungi is to occur, there must be a sustained supply of water to maintain the turgidity of the active cells. This does not necessarily mean that liberation is always most active under completely saturated conditions with no evaporation occurring. Indeed, it has been seen earlier that cup-fungi tend to 'puff' in response to sudden decrease in the humidity of the ambient air, and in *Sordaria fimicola* such a decrease leads to a sudden, but very temporary, increase in spore release (see p. 54). Nevertheless, over any prolonged period, spore discharge can be maintained only if the general loss by evaporation from the active cells is made good. It is the object of the present chapter to consider this water balance in relation to discharge.

In Ascomycetes a number of fleshy species belonging to such genera as *Peziza, Leotia, Helvella*, and *Morchella* have little power of withstanding drought and, like most agarics, the fruit-bodies can be produced and function only under sufficiently humid conditions. Drying destroys them. However, very many Ascomycetes, and particularly most Pyrenomycetes, are drought-enduring xerophytes. These, however, actually discharge their spores only under wet conditions. During dry periods they lose water and rapidly, but only temporarily, become inactive, quickly recovering and liberating their spores when wetted. This rapid response to wetting by rain largely explains the striking features of the damp airspora, for following rain, which combs many of the spores already present out of the air, there is soon discharged an abundance of the elongated septate ascospores so common in species of the large genus *Leptosphaeria* and its allies which occur on dead herbaceous plant material. Few precise observations seem to have been made of the ability of these xerophytic Ascomycetes to survive during long periods of dryness. However, when stromata of *Hypoxylon fuscum* were kept in a dessicated condition over anhydrous calcium chloride in sealed tubes in a refrigerator, they revived rapidly on wetting and discharged spores freely even after 18 months of storage. In nature, nevertheless, where repeated wetting and drying occur and where temperature fluctuates considerably, such a fungus as *H. fuscum* probably remains active for only a single season of not more than 6 months' duration.

The relationship of ascospore discharge to wetting has been studied in a number of pyrenomycetes. Gruenhagen (1945) showed that in *Hypoxylon pruinatum*, which causes a canker of poplar, a saturated atmosphere alone will not induce discharge; the stroma must be wetted. In another stromatal species, *Eutypa armeniacae* (a parasite of apricot trees in Australia), the picture is the same (Moller and Carter 1965). In this fungus abundant ascospore discharge follows rain, if this exceeds a daily value of 1·2 mm.

Venturia inequalis, a non-stromatal species that is responsible for apple scab, has been studied extensively (Hirst *et al.* 1955; Hirst and Stedman 1961, 1962). Perithecia develop on dead fallen leaves, maturation occurring in late spring. At this time of year there is a close correlation between periods of rainfall and high ascospore content of the orchard air (Fig. 11.1). Dew has little effect in bringing about discharge, but a light

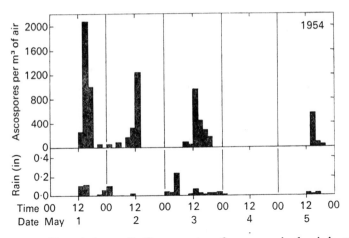

FIG. 11.1. *Venturia inaequalis.* Concentration of ascospores in the air in an apple orchard, and rainfall every two hours early in May. After Hirst *et al.* (1955).

rainfall of as little as 0·2 mm leads to abundant release of ascospores. The situation is very similar in *Guignardia citricarpa* which causes black-spot of citrus crops (e.g. orange). As in apple-scab the perithecia develop on the fallen leaves. Liberation of ascospores is dependent on rain, dew alone being ineffective (McOnie 1964). On the other hand in *Mycosphaerella pinoides*, foot rot of peas (*Pisum sativum*), which forms its perithecia on pea straw, although spore liberation still depends on wetting, this can be provided by dew alone (Carter 1963).

In spite of the fact that wetting seems necessary for discharge in so many Ascomycetes, a fruiting structure once fully wetted does not always continue at a high level of discharge. In *Ophiobolus graminis* Gregory and Stedman (1958) have shown that spore release begins shortly after wetting and the rate rises to a maximum within the hour. It then declines to zero after a few hours, and discharge will not recommence until drying has occurred followed by re-wetting (Fig. 11.2). Exactly the same

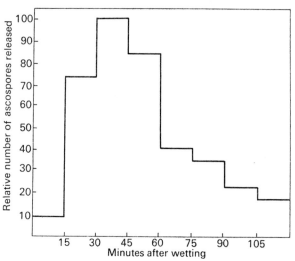

FIG. 11.2. *Ophiobolus graminis.* Rate of release of ascospores (spores trapped per 15 min as percentage of peak rate) in wind tunnel plotted against time. After Gregory and Stedman (1958).

behaviour has been reported in *Diatrype disciformis* (Ingold 1939), a common stromatal species on dead branches of beech (*Fagus sylvatica*).

In some pyrenomycetes it seems that a high level of discharge may occur when the wetted fungus begins to dry, and therefore shrinks. Lortie and Kuntz (1963) claim that this occurs in *Nectria galligena*, but the present author has been unable to demonstrate this in the closely-related coral-spot fungus, *Nectria cinnabarina*. However, essentially the same phenomenon seems to occur in *Sordaria fimicola*. We have seen (p. 54) that when an agar culture that has been under damp conditions is suddenly exposed to dry air, there is a dramatic, but very short-lived, rise in the rate of ascospore discharge. It has been suggested (Austin 1968) that this is due to incipient drying of the fleshy perithecium exerting pressure on the contents, and thus squeezing asci up to the ostiole where discharge occurs. Again, in lichens it is claimed that the violent ejection of ascospores from wetted fruit-bodies is greatest as they slowly dry (Ahmadjian 1967).

Although it must be emphasized that the drought-enduring type of xerophyte is the rule in Ascomycetes, there are a few examples in which discharge can be initiated and sustained, for a time at least, without the necessity for direct wetting.

A notable example is *Daldinia concentrica* which produces its large hemispherical perithecial stromata usually on dead branches and trunks of ash (*Fraxinus excelsior*) (Fig. 11.3). In Britain maturation occurs in

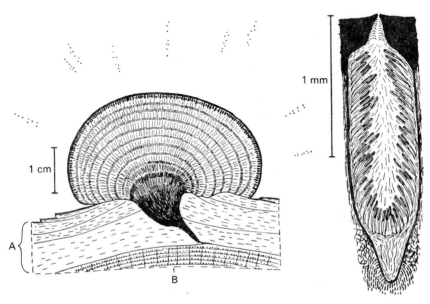

Fig. 11.3. *Daldinia concentrica. Left:* perithecial stroma on ash (*Fraxinus*) branch in sectional view with some spores just discharged in the air. The black dots just within the outer crust are perithecia: (A) bark; (B) wood. *Right:* a single perithecium.

May and spores are released at night throughout the summer. If a stroma, very early in its discharge activity, is detached and freely exposed indoors without any extraneous supply of water, spore discharge may continue for weeks. During this time the stroma, with its rigid crust, undergoes no appreciable change of volume, but its density declines from just over 1·0 to a value which may reach 0·3. Clearly during this period the water needed for spore discharge must have been obtained from the reserve in the stromatal tissue. Even if an isolated stroma is placed in a desiccator over anhydrous calcium chloride, spore discharge continues for many days (Ingold 1946*b*, 1960). It is by no means clear how water is transferred from the stromatal tissue to the perithecia, but sections show a three-dimensional network of veins that may be the channel of translocation.

A stroma attached to a tree seems to remain active much longer than a detached one. In Table 11.1 records are given of two isolated stromata

TABLE 11.1

Spore discharge in Daldinia

	Observations started	Start of discharge	End of discharge	Period of discharge (days)
I.	Detached stroma in laboratory			
	7 April	17 April	17 May	30
	10 April	27 April	23 May	26
II.	Stroma on tree-trunk			
	9 March	3 May	17 Sept.	138

that were brought indoors in early April. In these once discharge started it continued for about a month. Spore output was also followed at the same time from a parallel stroma attached to an ash tree. The set-up is shown in Fig. 11.4. Just above the stroma was a small wooden roof, as a protection from rain, and arranged below was a slide-holder in which a vertical glass-slide could be placed to catch the discharged spores. This slide was changed daily from early March to late September. Discharge started on 3 May and continued nightly until 17 September, a period of 138 days (see Table 11.1). This longer period of discharge from an attached specimen may well be associated with the gradual replenishment of the water reserve in the stroma by translocation from the tree trunk.

It is worth noticing that *Daldinia* is a fungus with a summer discharge period and this is possible because of its curious water-relations. Most other lignicolous Pyrenomycetes discharge their spores in autumn when conditions are more humid.

Another summer fungus with unusual water-relations is *Epichloe typhina*, causing 'choke' of grasses (Ingold 1948). Aerial shoots of infected grasses such as *Dactylis, Holcus, Brachypodium*, and *Agrostis* fail to flower. Instead, just above a node, a fungal stroma is produced closely wrapped around the living shoot. At an early stage it is whitish and produces abundant conidia, but in late July or August it turns orange-yellow becoming studded with closely-set perithecia. At this stage a transverse section of the affected region shows the fungal tissue intimately associated with the folded living grass leaves (Fig. 11.5). If a grass stalk attacked by 'choke' is cut with a razor (making the cut under water and thereafter keeping the cut end submerged) and brought into the laboratory, the course of spore liberation can be followed using the set-up illustrated in Fig. 11.5. The actual rate of spore output can be determined by placing a slide in a fixed position below the stroma and counting the spores deposited on a defined area in a given time. The rate is not constant; there

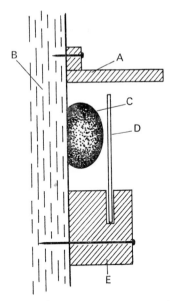

F IG. 11.4. *Daldinia concentrica.* Set-up used in study of spore discharge from a stroma on a standing tree: (A) roof; (B) ash trunk; (C) stroma; (D) glass slide; (E) slide-holder.

tends to be a day-time minimum and a night-time maximum. If in the evening, when the rate of spore liberation is high and rising, the stalk of the grass is severed above its water-supply, spore discharge falls to zero usually within the hour. Apparently the fungus derives the necessary water for discharge from the transpiration stream of its host, and is not dependent directly on wetting by rain.

Another example of an ascomycete in which spore discharge can continue in a dry atmosphere for some time without a sustained external supply of water is *Bulgaria inquinans* (Ingold 1959). The relatively large black apothecia, common on dead limbs and trunks of beech and oak in autumn, are somewhat like lumps of black india-rubber. The bulk of the thick apothecium, which may be 2–3 cm across and weigh as much as 10 g, consists of aqueous jelly representing a modification of the outer layers of the hyphal walls (Fig. 11.6). If a detached apothecium is hung upside down by a thread from a horizontal glass rod supported on the rim of a beaker, the spores deposited inside the beaker can be collected and counted at intervals. The gradual loss of water from the apothecium can at the same time be followed by weighing. It is found that in a dry atmosphere spore discharge can go on for several days, though at a declining rate. It ultimately ceases only when the apothecium is reduced to less than a quarter of its original weight. The actual picture of fall in the rate of spore release with decrease in weight is complicated by a diurnal

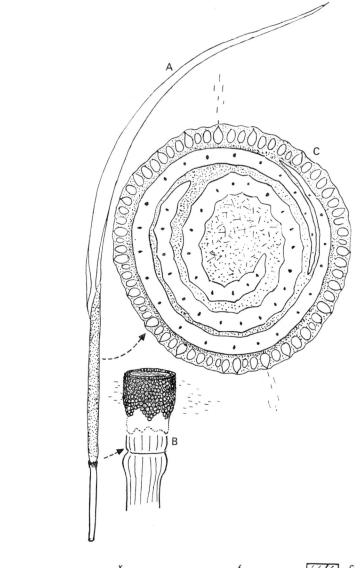

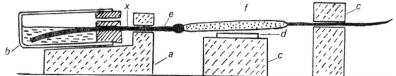

FIG. 11.5. *Epichloe typhina. Above :* shoot of *Dactylis* with perithecial stroma :
(B) details of region above node. (C) Transverse section showing stroma
with perithecia (fungal tissue dotted; leaves white with vascular bundles
as black dots). In B and C discharged spores are seen in the air around the
stroma. *Below :* method of measuring spore discharge: (*a*) support; (*b*)
specimen-tube with water; (*c*) wooden block; (*d*) glass slide; (*e*) stem of
grass; (*f*) stroma; (*x*) point of cutting to stop transpiration.

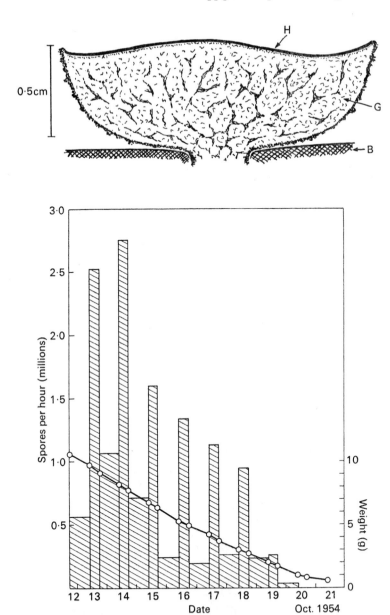

FIG. 11.6. *Bulgaria inquinans. Above:* vertical section of apothecium.
(H) hymenium; (G) gelatinous tissue; (B) bark. *Below:* spore liberation
from single drying apothecium. Continuous line relates to weight. Histo-
gram shows mean hourly spore liberation for day periods (10 a.m.–6 p.m.)
and for night periods (6 p.m.–10 a.m.).

rhythm in a specimen subjected to normal alternation of day and night (Fig. 11.6). Unlike the gelatinous Basidiomycetes, a dried and subsequently resoaked specimen of *Bulgaria* has little capacity for further spore discharge.

Sustained spore discharge under dry conditions over a period of several days without any outside supply of water is possible because of the available water stored in the jelly of the apothecium. It may, perhaps, be reasonable to suggest that wherever a relatively large amount of jelly is to be found in a fungal structure it represents an immediately available water reserve.

Another example of this principle is to be found in the stink-horn, *Phallus impudicus* (see p. 153). This is a fungus of summer and early autumn, which gradually stores water in the subterranean 'egg', mainly in the jelly of the middle peridium. When ripe the stipe within the 'egg' rapidly elongates carrying the cap with its stinking spore-slime above the substratum. The growth of the stipe, which happens with remarkable speed, requires a considerable supply of water which seems to come from the jelly.

We may now turn our attention to the huge group of Hymenomycetes and to the gelatinous Heterobasidiomycetes in all of which spore discharge occurs. The spores of the basidium are normally liberated in succession and it is only after the last one has been shot away that the basidium slowly collapses. Whatever may be the exact mechanism of discharge (see Chapter 5), it seems that the system can work only if the basidia retain their turgidity.

In those Basidiomycetes that actively shed their spores, as with Ascomycetes but more so, there are many species without any powers of resistance to drought. Most of the fleshy agarics are of this kind and their short-lived sporophores tend to be produced in autumn when the humidity of the air near the ground is high, and before the frosts of winter put a limit to their existence. But, as in Asocmycetes, there are many drought-enduring xerophytes that dry and cease to discharge their spores under conditions of drought. However, they remain alive, rapidly absorb water on being wetted by rain, and very soon afterwards begin to liberate spores again. To this group belong most of the corky and leathery polypores and other bracket fungi so common on wood, and also the gelatinous fungi such as *Auricularia*, *Tremella*, and *Calocera*.

Buller (1909) has drawn particular attention to *Schizophyllum commune* (Fig. 11.7). This is a small gill-bearing bracket fungus found all over the world and common in southern England on the dead trunks and branches of beech (*Fagus*). A curious feature of the genus is that the gills are split lengthwise. When the weather is damp and spore discharge is taking place, the gills look quite normal, but under dry conditions, when spore discharge stops and the sporophore shrivels, each gill separates from below upwards into halves that curl backwards covering up the whole hymenium.

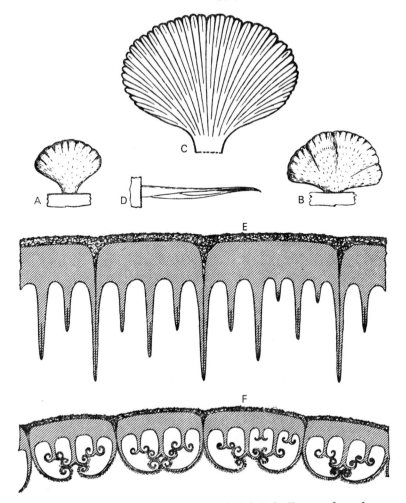

FIG. 11.7. *Schizophyllum commune.* (A and B) fruit-bodies seen from above growing on wood, × 1. (C and D) fruit-bodies seen from below and in section respectively, × 2. (E) section of fruit-body during wet weather. (F) section through fruit-body after drying. × 12. After Buller (1909).

Although this mechanism is remarkable and would seem to be associated with hymenium protection, its biological value may be doubted since many drought-resistant bracket fungi, such as *Polystictus versicolor* and *Stereum hirsutum*, seem to exist quite successfully without any structural elaboration of this kind. In *Schizophyllum*, as in *Polystictus* and *Stereum*, the hairy upper surface of the dry fruit-body quickly absorbs rain and the sporophore is soon in a condition to liberate spores again.

Early in the present century Buller sealed a number of dry sporophores of *Schizophyllum commune* in separate evacuated glass tubes. Some still remain, but tubes opened after over fifty years have been found capable of shedding spores soon after wetting (Ainsworth 1962). However, as with the drought-enduring Ascomycetes, it seems likely that in nature a sporophore lasts only for a single season.

There are a few Basidiomycetes in which the sporophores not only survive under very dry conditions but also continue to liberate spores in spite of drought. These are the large woody perennial polypores, particularly *Fomes fomentarius* and *Ganoderma applanatum*. The behaviour of these may be illustrated by a somewhat detailed consideration of *F. fomentarius*.

Buchwald and Hellmers (1946) obtained a cylindrical segment (about 60 cm high and 40 cm across) of a sycamore tree (*Acer pseudoplatanus*) bearing a large 2-year-old fruit-body about 32 cm in diameter. In October this was placed, with the sporophore in its natural orientation, on a table in the corridor of the Department of Plant Pathology of the Royal Veterinary and Agricultural College, Copenhagen where the temperature was about 18°C. On 24 March the sporophore started to shed spores and during 10 days there escaped visible, and photographably, white clouds, the average rate of spore liberation being 25 000 000 000 a day. Another small specimen on a trunk of *Quercus robur* brought into the Department in April shed spores from 17 May to 14 July when observations ceased. Previously Buchwald (1938) had made similar observations on a specimen growing on poplar.

The observations suggest that *Fomes* can derive the necessary water for spore liberation either from the tissues of the sporophore or from the tree trunk which is penetrated by the vegetative mycelium of the fungus. With a large active specimen of *Fomes fomentarius* detached in May from the beech trunk on which it was growing, the present author found that spore discharge, although at first copious, ceased after 2 days in the laboratory. This suggests that the sporophore itself does not have a sufficient water-reserve to sustain discharge and that the major reserve is in the tree. In this connection it should be realized that the water available in the wood is not merely that present as such, but also the water that can be liberated by the mycelium in the course of respiration by the oxidation of cellulose and other organic material. If the water in the wood is to be mobilized, an efficient conducting system must exist linking the fruit-body with this reserve.

The water relations of the very common *Ganoderma applanatum* (see p. 125) seem to be exactly like those of *F. fomentarius*. Isolated sporophores very soon cease to shed spores, but in the field sporophores can be seen liberating spores in rust-brown clouds even after a summer drought of over 6-weeks' duration.

At this stage in our knowledge we can only outline the factors that allow *Fomes* and *Ganoderma* to continue their spore liberation in very dry weather. First, general water-loss from the sporophore is probably cut down by the existence of a hard, woody upper surface from which evaporation may be relatively slow; secondly, the long narrow hymenial tubes probably ensure that the basidia are maintained in a saturated atmosphere even though the outside air may be relatively dry; thirdly, the very broad connection between the sporophore and the tree allows considerable hyphal connections with the water-supply in the tree trunk; and fourthly, the hyphae leading from the trunk to the hymenial tubes appear to follow a fairly straight course and, being apparently relatively free from cross-walls, seem structurally well suited to conduction. Very little is known about translocation in fungi, but there do not seem to be separate channels concerned with the movement of water and of organic food as in higher plants. In most fungi the intra-hyphal transport of water would appear to be quite inadequate to sustain a water-flow necessary to keep most fungal sporophores in a turgid condition during drought. However, in this connection the perennial polypores seem to be exceptional.

Not only must the turgidity of the hymenial elements be maintained if discharge is to continue in Hymenomycetes, but the air in immediate contact with the hymenium must also be saturated or nearly so. Very probably the essential significance of hymenia being displayed in tubes or on the closely packed gills or spines in hymenomycete sporophores is related to the necessity of maintaining high humidity for spore discharge.

Although high humidity must apparently be maintained at the surface of an hymenium of discharging basidia, liquid water has a ruinous effect. An hymenium of this kind is quite different from a discomycete one of asci and paraphyses that is generally uninjured by wetting. It is to be remembered that most toadstools and bracket fungi are so constructed that the hymenia are well protected from rain.

In fleshy agarics, which can hardly be regarded as xerophytes, a thick pileus may well represent a water reservoir, comparable with the stroma of *Daldinia*, which can, perhaps, supply water to the hymenia. Further, toadstools that have just started to dry can readily replenish their water reserves during rain. This capacity to absorb water easily has been studied by Buller (unpublished)* and his account of *Marasmius oreades* may be considered. However, it should be noted, that this species is one of the few fleshy agarics with xerophytic tendencies, since following a considerable degree of shrinkage as a result of drying, a sporophore is capable of subsequent recovery. Nevertheless, the process of raindrop absorption in this toadstool is probably the same as in many others which have much less capacity to withstand drying.

Normally, if a drop of water is placed on a pileus of *M. oreades* it remains for a minute or two and is then suddenly absorbed. A second

* Chapters of Buller's unpublished researches are deposited at the Royal Botanic Gardens, Kew, England.

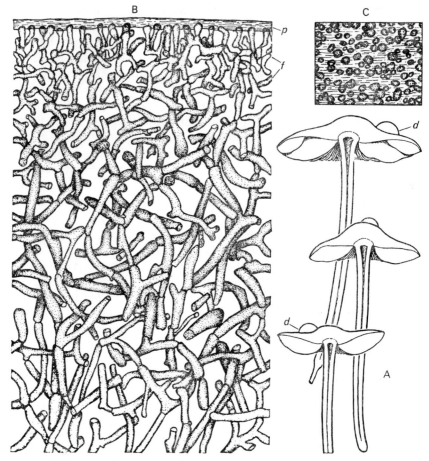

Fig. 11.8. *Marasmius oreades*. (A) three sporophores in vertical section with drops (*d*) placed on pileus. (B) vertical section of pileus: (*p*) pellicle; (*f*) fungal hyphae of tissue with air-spaces between. (C) surface section of pileus showing tips of hyphae (dark) projecting into mucilage of the pellicle. After Buller (unpublished).

drop placed on the same spot is similarly absorbed, the initial delay being reduced. However, ultimately no further drop can be absorbed, the specimen having become completely water-soaked. A fully expanded sporophore on a fine day has a continuous dry membrane forming the upper surface of the cap, and below it is a tissue of interwoven hyphae with air between (Fig. 11.8). When a drop is placed on the pileus, the membrane imbibes water and swells. Then, at a certain stage, the capillarity of the underlying tissue overcomes the imbibitional forces of the pellicle and water is sucked into the interhyphal spaces. Buller suggested that the advantage of the membrane is that being hard and dry under fine conditions it retards evaporation from the cap. However, when rain comes it swells rapidly and in the swollen condition allows the rapid passage of free water into the pileus.

The inter-hyphal capillary system of some fungi may also be important in translocating water from the substratum to hymenial surfaces. This happens, for example, in *Mitrula paludosa*. In this discomycete, with a club-shaped sporulating region borne on a stalk the base of which is submerged and attached to dead vegetation in very shallow water, the whole structure acts as a wick and water lost by evaporating from the aerial part is quickly made good by interhyphal capillary flow (Ingold 1954*b*). The same type of water replenishment no doubt also occurs in some agarics in which the base of the stipe is essentially submerged, for example in *Galerina paludosa* where the stipe rises from a bed of wet *Sphagnum*. Generally speaking, however, in fleshy fungi, although some translocation of water from the substratum through the stipe may occur, this is insufficient to supply the hymenium if the air is at all dry.

12. PERIODICITY

THE student of fungi is often faced with evidence of rhythms. Most are circadian*, but some are of longer period. Aerobiologists are familiar with annual cycles in the occurrence of spores in the air, a clear picture being available, for example, for the fall-out of *Cladosporium* conidia in Britain with a striking maximum in summer (Fig. 12.1). This probably

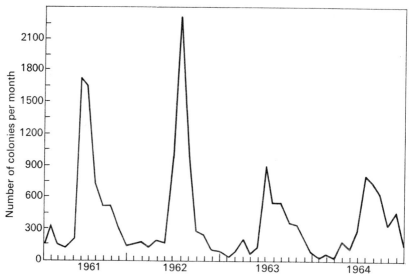

FIG. 12.1. *Cladosporium.* Monthly totals of colonies developing on agar in Petri dishes exposed for 10 min at Lechwith, Cardiff, S. Wales. After Harvey (1967).

reflects the production of fresh substrata, such as stubble, for saprophytic growth of the mould, combined with the stimulation of higher summer temperature. Another example, is the liberation of ascospores from stromata of *Hypoxylon rubiginosum* (Fig. 12.2), which goes on throughout the year but with a maximum in autumn (Hodgkiss and Harvey, 1969).

* The term 'circadian' for a rhythm in which one whole cycle is completed in about 24 hours has come to stay. Some would like to use 'diurnal' instead, but this is more precisely valid as the opposite of 'nocturnal'.

Rhythms are normally connected with the periodicity of external conditions, but there are a few exceptions. One such is to be found in the
discharge of glebal masses from a fruiting culture of *Sphaerobolus stellatus*
(Alasoadura 1963). When this is grown on oatmeal agar under continuous
light of constant intensity at 20°C, discharge, which may go on for several
months if the medium is deep enough, has a periodicity with 10–12 days

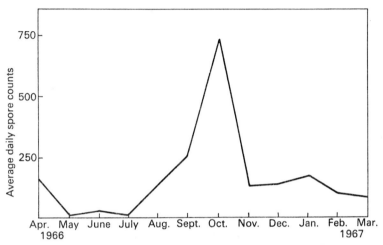

FIG. 12.2. *Hypoxylon rubiginosum.* Seasonal variations shown by average
daily spore-counts for each month from April 1966 to March 1967. Hirst
spore-trap used with nearby stromata-bearing logs freely exposed on soil
trays outside to natural conditions. After Hodgkiss and Harvey (1969).

between maxima (Fig. 12.3). This interval agrees closely with the time
from the first visible initiation of sporophores to maturity, and suggests
that existing fruit-bodies inhibit the development of new ones until they
have themselves discharged their glebal masses. This view is strongly
supported by experimental evidence (Ingold and Nawaz 1967).

Circadian rhythms in spore liberation are to be expected because the
principal factors that influence take-off themselves show a daily periodicity
to a greater or lesser extent. Light and temperature both have a day-time
maximum, whereas humidity is generally highest at night. In a more
general manner, being apparent only by averaging over long periods, wind
velocity has a day-time peak. Again rainfall, which can profoundly affect
spore liberation both directly and indirectly, may on average show a
circadian periodicity, but this varies from one part of the world to another.

Aerobiologists who study the concentration of spores in the air throughout the 24-hour period using volumetric traps almost invariably record a
periodicity for each type of spore. Usually the results are reported in the
form of a graph, each point being a geometric mean (based on determinations for many consecutive days) of the concentration of spores in the air

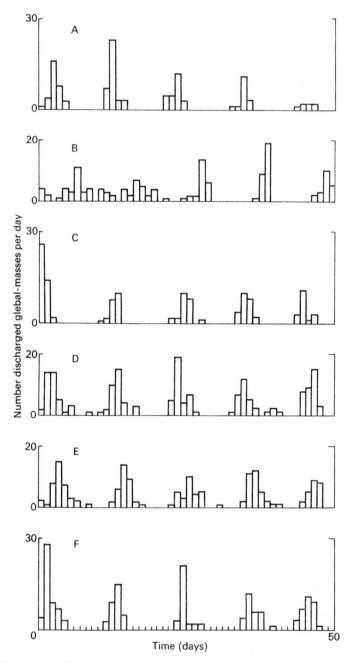

FIG. 12.3. *Sphaerobolus.* Daily glebal-mass discharge (starting from first day of discharge) in six parallel cultures in continuous light (1000 lx) at 20°C. Time before discharge started varied from 20 to 24 days. After Alasoadura (1963).

at a particular time of day, the successive points on the graph being at hourly or 2-hourly intervals (Fig. 12.4). A number of patterns can be recognized (Gregory 1961) according to the time of day when the spores are most plentiful in the air. For example *Sporobolomyces*, *Tilletiopsis*, and the basidiospores of rusts show nocturnal peaks; a few species (e.g. *Peronospora tabacina* and *Deightoniella torulosa*) have a clear maximum shortly after dawn; others such as *Phytophthora infestans* have the peak later in the morning, while still others have their maxima in the afternoon (e.g. rust urediospores, smut spores and the conidia of *Cladosporium* and *Alternaria*).

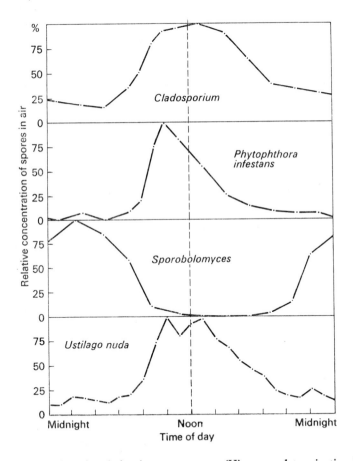

FIG. 12.4. Diurnal variation in spore content (Hirst trap determinations) expressed as percentage of peak geometric mean concentration. Curves for *Cladosporium* and *Phytophthora* based on observations on agricultural land at Rothamsted (Hirst 1953) for *Sporobolomyces* curve based on observations at Thorney Island, Chicester Harbour (Gregory and Sreeramulu 1958); *Ustilago* curve based on observations in infected barley field at Silwood Park, Berkshire (Sreeramulu 1962).

It must be emphasized that, although the mean pattern revealed in a study of a constituent of the air spora over the 24-hour period is probably often related to periodicity of spore liberation, this is not necessarily so. A diurnal rhythm of spores in the air might be observed unrelated to rhythmic take-off. Let us consider, say, the apothecium of a discomycete growing on the ground and discharging spores at a uniform rate to a height of a centimetre or two. During the night in fair weather there is often a boundary layer of still or laminar air several centimetres deep in contact with the ground. Spores discharged into this would soon fall to the ground and would not be caught by a Hirst trap operating at the standard height of 1 m. During the day, with the onset of turbulent conditions and the reduction of the boundary layer to a bare millimetre or less, discharged spores would be brought into the trapping region. Thus a periodicity might be recorded having no connection with periodic spore liberation.

Some rhythms in elements of the air spora are clearly related to periodicity of environmental factors that influence the release of spores. One of these is the rapid drying of the air near the ground shortly after sunrise. Meredith (1961, 1962b), working in banana plantations on *Deightoniella torulosa* (causing 'speckle' of the fruit) and *Cordana musae* (producing a minor leaf-spot), has found with both fungi a steep rise in the concentration of conidia in the air just after dawn associated with a sharp decline in the relative humidity of the air (Fig. 12.5). This can be related to the mechanism of spore take-off in both species, involving water-rupture as a result of drying in the cells of the conidiophore. A very similar periodicity has been observed for spores of *Peronospora tabacina* in the air above tobacco fields affected by 'blue-mould'. Here the early morning maximum is no doubt connected with the twirling action of the conidiophores on drying, which flings off the spores (Waggoner and Taylor 1958). Again the Kansas aerobiologists (Pady *et al.* 1968) have recorded a periodicity of aeciospore liberation in *Gymnosporangium juniperi-virginianae*, growing on apple leaves, with a sharp morning maximum apparently associated with drying that causes the emergent peridial filaments of the aecium to curve backwards so that the spores can escape. On the other hand, in aecia of species of *Uromyces* and *Puccinia*, without a hygroscopic peridium, discharge is confined to damp periods when the turgor of the aeciospores is at a maximum (Fig. 12.6). In these rusts aeciospores tend to be discharged at night but this nocturnal rhythm may be upset when daytime rains wet the leaves (Kramer *et al.* 1968).

Reference has already been made to the fact that *Cladosporium* in the air at Rothamsted in the United Kingdom tends to have a maximum around noon (Fig. 12.4). However, it has been found in Kansas, U.S.A. that a regular and very emphatic maximum occurs in the early morning. The difference might be due to the considerable differences in the two localities, but it is more likely to be related to the fact that in one region the general air-spora was being sampled and spores were being trapped that had been liberated over a wide area and many of them had probably been in the air for a considerable time, whereas the Kansas workers were trapping spores within a few centimetres of the point of liberation. Thus essentially spore liberation was being studied and the morning peak was related to a rapid fall in the humidity of the air combined with a circadian ripening of the spores (Pady *et al.* 1969).

In an intensive study Sreeramulu (1962) found that the maximum concentration of chlamydospores of loose smut of barley (*Ustilago nuda*)

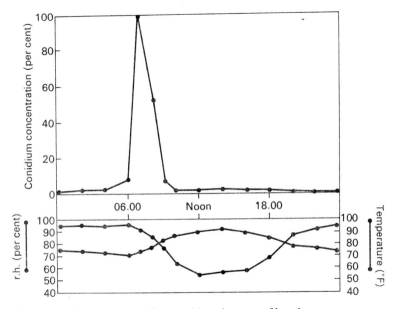

Fig. 12.5. *Cordana musae. Above:* arithmetic mean of hourly spore concentration (based on counts on over 100 days) as percentage of peak value. *Below:* corresponding mean hourly value for temperature and r.h. in the banana plantation. After Meredith (1962*b*).

in the air at a height of 90 cm in the centre of a circular infected barley plot occurred, on the average, during the midday hours (Fig. 12.4). This maximum appeared to be closely correlated with wind velocity, which tends to be at its highest around noon. It is perhaps of significance to remark that during the hours around noon not only does the wind tend to be strongest, but also to have its greatest degree of gustiness, an important factor where stem vibrations or leaf flutter assist spore liberation. Barley plants infected with *U. nuda* are normally a little taller than healthy ones, so that the smutted heads stand slightly above the general level.

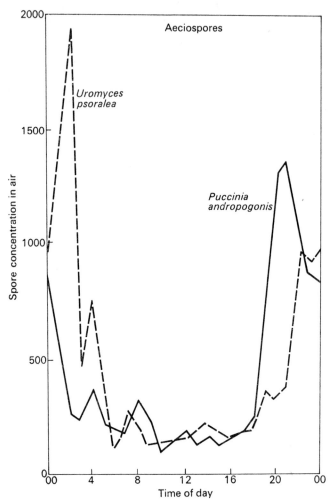

FIG. 12.6. *Puccinia andropogonis* and *Uromyces psoraleae*. Periodicity of aeciospores (as concentration per ft³) in air sucked from the close neighbourhood of diseased leaves. *P. andropogonis*: mean of 22 rainless days; *U. psoraleae*: mean of 31 rainless days. Based on figures by Kramer *et al.* (1968).

Further, diseased heads are relatively light and remain upright, while normal stalks have heavier heads that bend over away from the wind (Fig. 12.7). Another point to bear in mind is that the smut spores of an infected head mature more or less together, but their liberation is spread over a prolonged period. There is, apparently, no circadian maturation of spores, so that the periodicity observed in the field seems to relate entirely to the rhythm of the external conditions affecting take-off itself.

A rather similar picture is presented by *Monilinia laxa*, the causal organism of blossom wilt, spur blight, and brown rot in stone fruits. The conidia form dry powdery masses and are blown off by wind. It has been found that in the air of an affected apricot orchard in the Sacramento

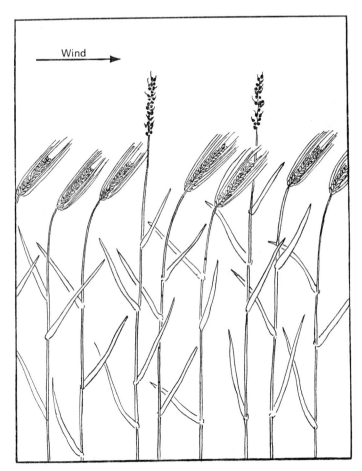

FIG. 12.7. *Ustilago nuda*. Diagram of a barley field with the smutted inflorescences of two infected stalks projecting above the healthy heads bent over in the wind indicated by the arrow.

Valley of California, the concentration of conidia is correlated directly with the wind velocity. The maximum spore concentration in the air is just before sunset and this is highly correlated with the wind velocity in the Sacramento Valley which reaches its maximum at this time of day (Fig. 12.8).

However, sometimes, although the actual take-off may be conditioned by midday winds, the production of a daily crop of spores, ripe for dispersal, may be under the control of other factors with a circadian rhythm such as light or temperature or both. Urediospores of many rusts seem to show this type of behaviour. The peak concentration of urediospores in the air tends, as with smut spores, to be around midday (e.g. Sreeramulu 1959) and appears to be associated with wind velocity. However,

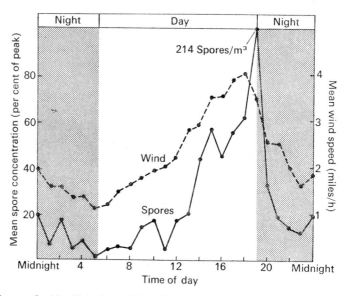

FIG. 12.8. *Monilinia laxa*. Mean hourly concentration over period 1–27 July as percentage of maximum of conidia in air within an affected apricot orchard in California 2 m above ground and corresponding values of mean wind velocity within the orchard during the same period. Time of day is Pacific standard time. Slightly modified from Corbin, Ogawa and Schultz (1968).

the actual ripening of urediospores is also diurnal (Kramer and Pady 1966) and is conditioned by both temperature and light (Smith 1966).

It is of some interest to consider in general terms periodicity in dispersal of spores in the rusts (Uredinales). In contrast to the diurnal pattern of urediospore liberation (Fig. 12.9), the discharge of aeciospores, usually involving turgor and therefore requiring high humidity, tends to be nocturnal, especially during periods of little rainfall (Kramer *et al.* 1968). However, *Gymnosporangium juniperi-virginianae* is an exception since its hygroscopic peridium allows aeciospores to escape only when the relative humidity of the air is sufficiently low. Thus the peak of liberation is associated with the sharp decrease in the humidity of the air after dawn (see p. 134). For the production and liberation of the thin-walled basidiospores (sporidia) of rusts damp conditions are also necessary and again discharge tends to occur at night (Fig. 12.10) when humidity is at its highest (e.g. Carter and Banyer 1964).

Probably the maximum in the small hours before dawn for ballisto-spores of mirror-picture yeasts (*Sporobolomyces* and *Tilletiopsis*) in the air spora is largely associated with the high humidity necessary for sustained spore discharge (Zoberi 1964), but light may also play a part (Bridger 1967).

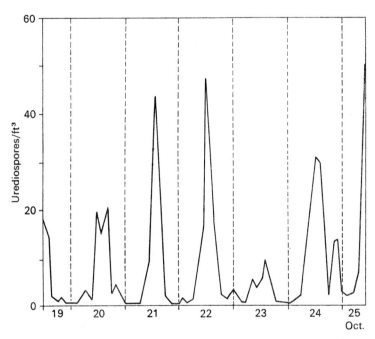

FIG. 12.9. *Puccinia recondita*. Field record of urediospores in air in Kansas, U.S.A. Sampled close to ground amongst infected leaves in a wheat plot. Midnight positions shown by dashed lines. After Kramer and Pady (1966).

We may now pass on to a consideration of circadian periodicity of spore liberation revealed by laboratory studies where there has been close control of the environment.

Considerable attention has been given to rhythms that are associated with light. We have seen (Ingold and Dring 1957) that in *Sordaria fimicola*, provided the overall water-supply is assured, a very pronounced diurnal rhythm in spore discharge occurs, under conditions of 12 hours' light and 12 hours' darkness in each 24-hour period at 20°C, with the day-time maximum reached long before the close of each light period (Fig. 2.37). A similar periodicity (Fig. 12.11) has been discovered in a wide

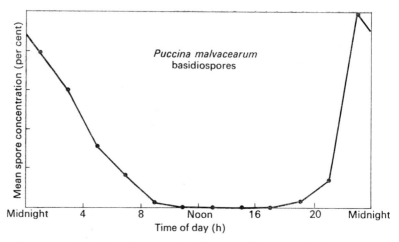

FIG. 12.10. *Puccinia malvacearum* basidiospores. Concentration (as percentage of peak geometric mean) throughout day over period of 21 days in air near rusted *Malva parviflora* at Adelaide, Australia. After Carter and Banyer (1964).

range of sordariaceous fungi (Walkey and Harvey 1966, Callaghan 1962).

On the other hand, in *Apiosordaria verruculosa* (Ingold and Marshall 1963) with the same light: dark regime, rather more spores are usually discharged in the dark periods (Fig. 12.12). This, however, is not apparently the result of inhibition by light. Indeed, discharge is stimulated by change from darkness to light and inhibited by the reverse transition, but at 20°C there is an interval of 8–12 h between the reception of the stimulus and maximum response.

A number of pyrenomycetes are, however, essentially nocturnal (Fig. 12.13). This is particularly true of species of the Xylariaceae (e.g. *Hypoxylon fuscum, Daldinia concentrica, Xylaria longipes*), but other species outside this family are also nocturnal, for example, *Hypocrea gelatinosa* (Kramer and Pady 1968) and *Bombardia fasciculata* (Pady and Kramer 1969).

In flask-fungi (Pyrenomycetes) it seems that species tend to be either diurnal or nocturnal, always provided that water-supply is not limiting. We have noted a tendency for Sordariaceae to be diurnal and Xylariaceae nocturnal. However, the difference in spore discharge does not always follow taxonomic lines. Thus amongst bitunicate Ascomycetes *Sporormia intermedia* is diurnal, but *Melanomma pulvis-pyrius* is nocturnal. Again, *Nectria coccinia* is nocturnal but *N. cinnabarina* liberates its ascospores by day. To some extent, perhaps, an ecological distinction is more valid. Certainly coprophilous fungi generally tend towards diurnal discharge, while lignicolous fungi, so far as Ascomycetes are concerned, are most often nocturnal.

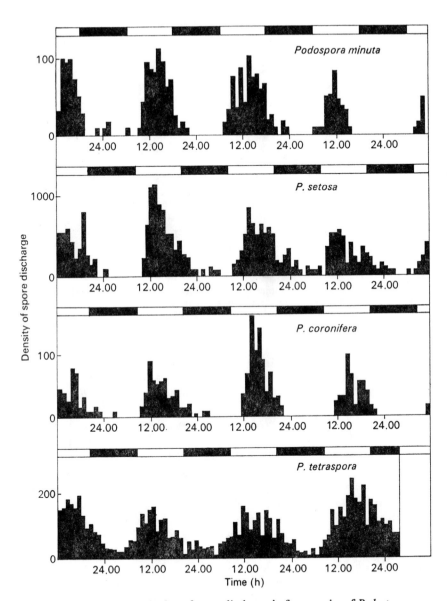

FIG. 12.11. Diurnal rhythm of spore discharge in four species of *Podospora*.
After Walkey and Harvey (1966).

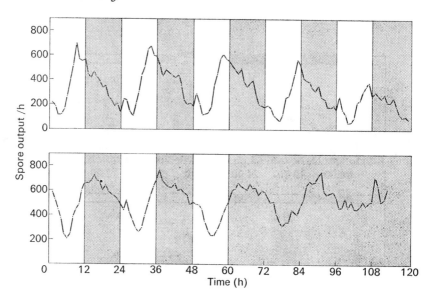

FIG. 12.12. *Apiosordaria verruculosa.* Course of spore discharge in two parallel cultures at 20°C subjected to a regime of 12 h light: 12 h darkness. The second culture (lower graph) was finally placed in continuous darkness. Dark periods stippled.

Coprophilous fungi, irrespective of taxonomy, exhibit diurnal periodicity of discharge. This happens in *Ascobolus* spp. and related cup-fungi, in *Sordaria* spp. and their allies, in *Pilobolus* spp. and *Basidiobolus ranarum* (on frog's excrement), and in *Sphaerobolus stellatus*, which is both coprophilous and lignicolous. In all these the direction of discharge is conditioned by light, a feature that tends to get the spores onto the surrounding grass. Thus the rhythms can be related to periods when phototropic response is possible and seems to have selective advantage. It is more difficult to suggest what may be the value to lignicolous fungi of nocturnal spore liberation.

Again the timing of liberation may possibly be related to the type of spore involved. Thus in rusts the rather delicate basidiospores are generally liberated at night when they are not liable to injury by light and may be less liable to undue dessication since the air tends to be damper. However, the much larger and more resistant urediospores mainly become airborne around midday. Their thick and sometimes pigmented walls apparently protect them from radiation, and being larger they are more dependent on day-time turbulence for their effective dispersal. However, this argument is weakened by the fact that aeciospores, which in size and general structure closely resemble urediospores, are, as we have just seen, usually discharged at night.

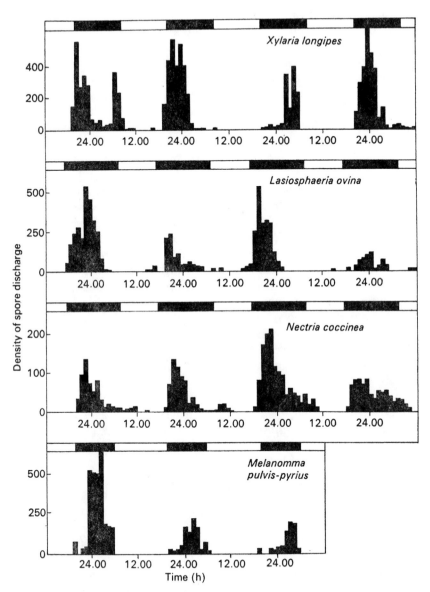

FIG. 12.13. Periodicity of spore discharge in 'nocturnal' species of pyrenomycetes. After Walkey and Harvey (1966).

Returning to the consideration of nocturnal flask-fungi, in *Hypoxylon fuscum* at least, periodicity is related to the direct and immediate inhibitory action of light on discharge even at low intensities (Ingold and Marshall 1963). In *Daldinia concentrica*, however, the nocturnal rhythm is to some extent endogenous, continuing for a time after the circadian periodicity of illumination has ceased (Ingold and Cox 1955). When a stroma is subjected to 12 hours' light and 12 hours' darkness each day, nearly all spores are shed in the dark periods and this is also true of a specimen under natural conditions on an ash tree with the normal alternation of day and night. When a specimen is transferred to darkness at constant temperature and humidity, periodic spore liberation continues for a number of days, as many as twelve, but thereafter, although spores continue to be discharged, the rhythm is lost (Fig. 12.14). However, on return to the original periodic conditions, the old rhythm is immediately renewed. On transfer from natural conditions to continuous light, a stroma also retains its rhythm of spore discharge, but the periodicity is lost in 2 or 3 days, much more rapidly than in the dark. By subjecting a specimen to four 6-hour periods each day with light alternating with darkness, a periodicity is soon produced with two maxima, one in each dark period and two minima corresponding with the periods of illumination in the course of 24 hours. When, however, a stroma so conditioned is placed in darkness, only the original night-time peak is 'remembered'. Attempts to produce a peak of discharge every other day, by a regime of 24 hours' light followed by 24 hours' darkness, have failed. With this regime the fungus still has its peaks of maximum discharge with approximately a day between.

Although in Ascomycetes a light-conditioned rhythm of spore discharge appears to be common, it is apparently rare in Basidiomycetes. An example has, however, been described by Carpenter (1949) in *Pellicularia filamentosa* causing a leaf spot of *Hevea* rubber. This fungus in its spore discharge seems to resemble *Daldinia* particularly closely. With the normal alternation of day and night in Peru nearly all the spores are liberated in the dark, mainly before midnight. The periodicity of discharge seems to be endogenous for its goes on for several days in either continuous light or continuous darkness.

In many ways the story of periodicity in a culture of *Pilobolus* resembles that of a stroma of *Daldinia* except that the circadian rhythm in *Pilobolus* involves a peak in discharge-rate during the day-time. It is important in any discussion of *Pilobolus* to emphasize that not all species behave in exactly the same way. In the two most exhaustive studies of this subject, by Schmidle (1951) and by Uebelmesser (1954), *Pilobolus sphaerosporus* has been used, although in a few of her experiments Uebelmesser used *P. crystallinus* as well. *P. sphaerosporus* appears to be much easier to grow in the laboratory than other species. Further, in this, unlike other species,

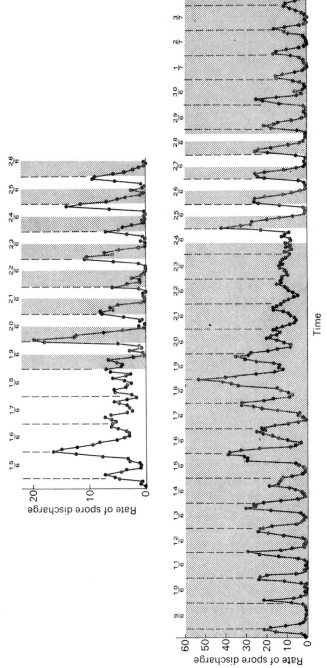

FIG. 12.14. *Daldinia concentrica*. Rate of spore discharge plotted against time. Upper graph: stroma in continuous light for 4½ days and thereafter given a daily regime of 12 h light: 12 h darkness. Lower graph: different stroma in darkness for 16 days; then with daily regime of 12 h light: 12 h darkness for 5 days; then again in dark. Both stromata previously under normal field conditions. Temperature 21°C, light *c*. 1000 lx. Dark periods dotted. Vertical interrupted lines show midnight positions. To convert rate of discharge into spores trapped per 2-h period the ordinates must be multiplied by 10⁵. After Ingold (1960).

sporangiophores can develop in the complete absence of light, although sporulation is greatly stimulated by illumination.

With an alternation of 12 hours' light and 12 hours' darkness, humidity and temperature being kept constant, both German workers found a very pronounced peak of sporangial discharge about the middle of each light period and this periodicity was retained for several days following transfer to continuous darkness (Fig. 12.15). Uebelmesser, as did Schmidle, investigated the effect on the rate of discharge of a considerable range of light: dark regimes of varying length. For example, with 6 hours' light alternating with 6 hours' darkness, peaks of discharge developed near the middle of each light period. On transfer to continuous darkness, however, although in the first 24 hours two peaks occurred at 12-hour intervals, following that the peaks were 24 hours apart. The same picture emerged with a regime of 4 hours' light: 4 hours' darkness. Peaks developed in the short light periods and, on transfer to uninterrupted darkness, these remained and were still 8 hours apart, but after 24 hours only a single daily peak persisted.

As with *Daldinia*, it does not seem possible to induce *Pilobolus* to adopt a 48-hour rhythm. Thus Uebelmesser found that whereas a regime of 1 hour's light: 23 hours' darkness gave peaks of discharge activity a day apart, so did one involving 1 hour's light: 47 hours' darkness.

In *P. sphaerosporus* a periodicity of temperature can also induce a rhythm in discharge of sporangia. Uebelmesser, using a dark-grown culture maintained at a 'ground' temperature of 15°C, gave it a daily treatment of 1 hour at 25°C. This led to a diurnal rhythm with a peak about half-way between each pair of warm periods. Further, in a similar culture but maintained at 25°C and given a daily hour at 15°C, the same type of rhythm developed with the peaks this time midway between the cold periods. In both cases the rhythm, having become established as a result of 2 or 3 days' treatment, persisted for at least 4 days at the uninterrupted 'ground' temperature. Again, as with light, a rhythm with two peaks in the 24-hour period could be induced by two separate hours at a high temperature (30°C) spaced 12 hours apart (Fig. 12.16). Following this if the culture was kept at a steady 'ground' temperature (20°C), the rhythm persisted with two peaks in the first 24 hours, but thereafter there was only one a day.

In both *Daldinia* and *Pilobolus* there seems to be a biological clock. In *Pilobolus* this can be set, by either light or temperature, to a 24-hour period, and once set the clock keeps time for several days without the necessity for outside resetting.

How commonly endogenous circadian rhythms occur in relation to spore liberation in fungi is uncertain. In extensive experiments with *Sordaria fimicola*, the present writer and his co-workers originally considered that there was no endogenous rhythm in that species. Later work

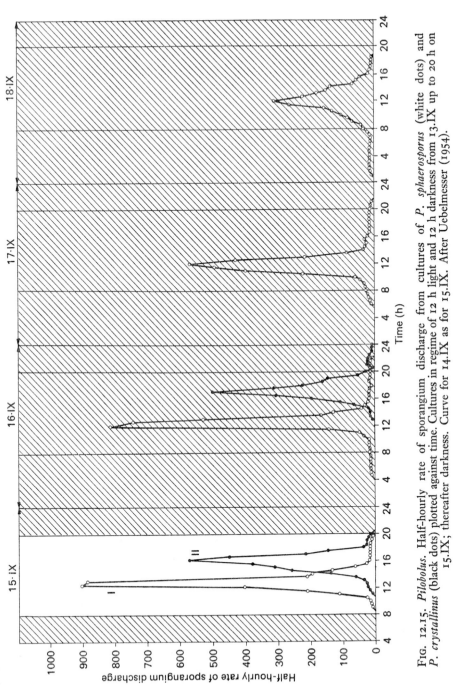

Fig. 12.15. *Pilobolus*. Half-hourly rate of sporangium discharge from cultures of *P. sphaerosporus* (white dots) and *P. crystallinus* (black dots) plotted against time. Cultures in regime of 12 h light and 12 h darkness from 13.IX up to 20 h on 15.IX; thereafter darkness. Curve for 14.IX as for 15.IX. After Uebelmesser (1954).

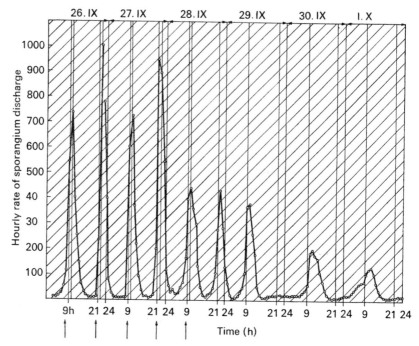

Fɪɢ. 12.16. *Pilobolus sphaerosporus.* Hourly rate of sporangium discharge
from a culture kept in dark at a general temperature of 20°C raised to 30°C
for 1 h twice in 24 h (at 9 h and 21 h). Last high temperature treatment at
9 h on 28.IX. After Uebelmesser (1954).

by Austin (1968*b*), however, has indicated that such a rhythm does exist,
but only for part of the life of a perithecium, and is easily obscured by
the much stronger periodicity induced by alternation of light and dark
periods in each 24 hours. It is but a step from this condition to that
observed by Bridger (1967) in *Sporobolomyces roseus* where no endogenous rhythm was discovered, but while a nocturnal periodicity was
immediately induced by 12 hours' light: 12 hours' darkness daily, no
regular rhythm resulted when cultures were subjected to alternating
24-hour periods of light and darkness. The fungus has, apparently, an
inherent ability to assume a circadian rhythm in response to an appropriate periodicity of the light regime.

Sphaerobolus stellatus is another fungus exhibiting, in the discharge of
its glebal masses, a diurnal periodicity clearly related to alternation of
light and darkness. In this gasteromycete a soft adhesive spore-containing
glebal mass, about the size of a small mustard seed, is catapulted from the
ripe opened sporophore to a distance of up to several metres. Unlike most
Gasteromycetes, it can be grown easily in pure culture in the laboratory.
On an appropriate medium and with other environmental conditions

properly adjusted, it fruits abundantly (Alasoadura 1963). Initiation of fruit-bodies takes place only if the light intensity is sufficiently high and then only if the temperature is below a limiting value of approximately 25°C. Not only is light absolutely necessary for the initiation and early growth of sporophores, but light also strongly stimulates their further development, although the minimum effective intensity steadily falls. At 20°C the total period of sporophore development is about 14 days, from the time a dark-grown culture is first subjected to light, and sporophores are presumably initiated, until finally the mature spherical fruit-body opens in a stellate manner about 5 hours before its glebal mass is

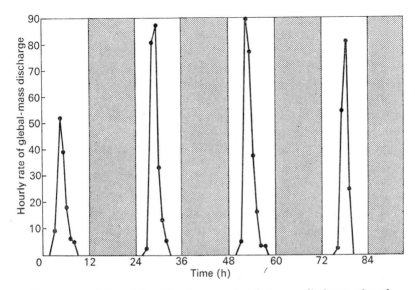

FIG. 12.17. *Sphaerobolus*. Hourly rate of glebal-mass discharge plotted against time in a regime of 12 h light: 12 h darkness (shaded periods). After Friederichsen and Engel (1960).

discharged. A remarkable feature of the story of development is that while initiation and early sporophore development depend on visible light of shorter wavelength (below 500 nm), the later stages are especially stimulated by light of longer wavelength (above 550 nm) and blue light has then, indeed, a retarding influence. However, the stage, covered by the final 24–28 hours before discharge, is completely independent of light.

A study of periodic discharge has been made by Friederichsen and Engel (1960) and their findings have been confirmed and extended by workers in my laboratory (Alasoadura 1963, Ingold and Nawaz 1967). Under conditions of 12 hours' light and 12 hours' darkness each day, discharge is entirely in the light periods (Fig. 12.17) provided the ambient temperature is around 20°C, but with alternating 24-hour periods of light and darkness, discharge is almost limited to the dark periods. If, after the regime in Fig. 12.17 the culture had been transferred to continuous darkness at 84 hours, there would have been another peak in the 12-hour period 96–108 hours but no discharge in the following few days. This additional peak is not the result of an endogenous rhythm that soon dies out, but is a consequence of the fact that the final stages of development are not light-sensitive and discharge is conditioned by illumination received about 24 hours earlier. This also explains the occurrence of discharge in the dark periods when the regime consists of 24 hours' light and 24 hours' darkness.

The duration of the final light-insensitive period depends on temperature, and at 10°C it is increased from 26 to about 39 hours. It is therefore, not surprising that at this lower temperature a culture of *Sphaerobolus*, given a 12-hour: 12-hour light: dark regime, discharges its glebal masses mainly in the dark periods, not in the light ones as at 20°C.

Although in *Sphaerobolus* periodicity of discharge can be firmly related to the rhythm of light treatment with a lag of about a day at 20°C, there is also some evidence for the existence of an endogenous rhythm. When a fruiting culture in continuous light is transferred to darkness, discharge ceases after just over a day. Then, usually 4–6 days later, some further discharge occurs for a few days, and it is claimed that this tends to show a circadian rhythm (Engel and Friederichsen 1964).

In *Sphaerobolus* diurnal discharge at temperatures around 20°C and with an alternation of light and darkness similar to that occurring under natural conditions can clearly be related to stimulation during the light periods, development being temporarily arrested during the dark ones. In continuous light the course of discharge is uninterrupted. In *Pyricularia oryzae*, however, both light and dark periods appear to be necessary if the spores are to be set free.

P. oryzae causes 'blast' of rice. Spores are set free from the conidiophore only if the air is saturated or nearly so, and it seems that an active discharge mechanism is probably involved (Ingold 1964). Barksdale and Asai (1961) have studied the release of spores from infected rice leaves. The process is nocturnal under conditions resembling the natural alternation of night and day. When, however, *P. oryzae* is subjected to continuous darkness, discharge ceases after about 12 hours and is not resumed in darkness until there has been a prior period of illumination (Fig. 12.18). Similarly, spore liberation ceases in continuous light. It would thus seem that both dark and light treatments are needed if spores are to mature and be discharged.

In fungi that show a diurnal rhythm of discharge related to light, the maximum is commonly near the middle of each light period and the rate begins to decline before the onset of the next dark one. This is the pattern in most sordariaceous fungi, in *Pilobolus*, and in *Sphaerobolus*. In *Basidiobolus ranarum* (Callaghan 1969), a diurnal species with no sign of an endogenous rhythm, the picture is in general the same, but there tend to be two peaks in the light periods (Fig. 12.19). This may be due to the existence of two critical stages in conidiophore formation. It seems that

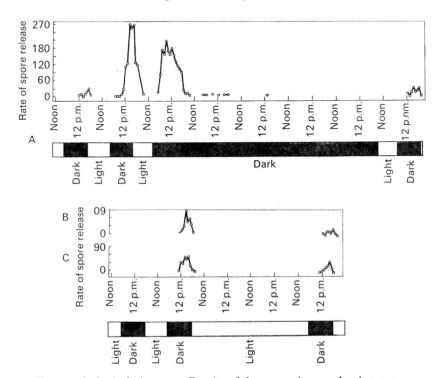

FIG. 12.18. *Pyricularia oryzae*. Results of three experiments showing rate of spore discharge plotted against time. The light-dark regime is indicated. Spores are set free in dark periods following light ones. After Barksdale and Asai (1961).

early development may, up to a point, occur in darkness, although it is also stimulated by light. Later, light is essential. Each day, on transfer from dark to light, the accumulated young conidiophores, that have developed up to but not beyond a certain point, very rapidly mature and discharge their spores to give the first peak, the later one resulting from discharge from conidiophores that have developed from the beginning under the stimulatory influence of light. This explanation is supported by the behaviour of cultures which, following inoculation, are kept in darkness for a few days before transfer to continuous light, when the rate of discharge rises rapidly from zero to a high peak, and then quickly

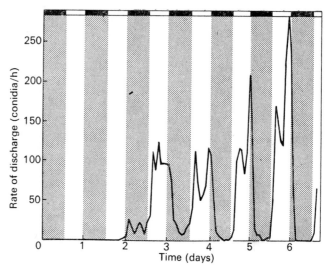

FIG. 12.19. *Basidiobolus ranarum.* (A) rate of spore discharge from a culture subjected to a daily alternation of 12 h: 12 h light and darkness. Time from first inoculation. After Callaghan (1969).

declines to a low level from which it gradually recovers over a period of several days. The course of discharge from a culture kept in light from the time of inoculation is strikingly different (Fig. 12.20).

A circadian rhythm of discharge under a regime of 12 hours' light: 12 hours' darkness is not always associated with a difference in basic rate of discharge in continuous light as compared with continuous darkness, although this is often so (e.g. in *Basidiobolus ranarum* or *Sordaria fimicola*). In *Sporobolomyces roseus* and in *Conidiobolus coronatus* the pattern of discharge from parallel cultures in light and darkness over a period of several days is essentially similar, both qualitatively and quantitatively. Nevertheless under a regime involving 12 hours' light: 12 hours' darkness each day, a very marked circadian rhythm is established; diurnal in *Conidiobolus*, but essentially nocturnal in *Sporobolomyces*.

The state of affairs in powdery mildews (Erysiphales) is interesting. Pady and his colleagues (1969) have studied the liberation of conidia from *Erysiphe* spp. on infected leaves in growth chambers maintained at 90 per cent r.h. and 21°C with an alternation of light and darkness (12 h: 12 h, or 8 h: 16 h) each day. In *E. polygoni*, in which, as already noted by Yarwood (1936), each conidiophore produces and releases only one conidium daily, this is always set free in the light period (Fig. 12.21). In *E. cichoracearum*, however, up to ten are abstracted daily, being produced throughout the 24-hour period; but their release is largely confined to the light period. In *E. graminis*, in striking contrast to the two

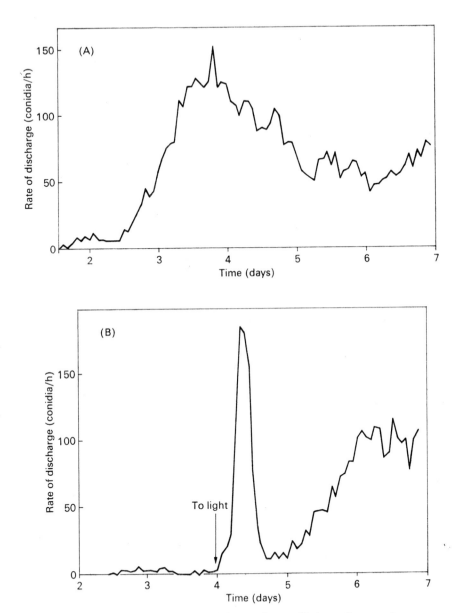

FIG. 12.20. *Basidiobolus ranarum.* (A) rate of spore discharge from a culture in continuous light (1700 lx). (B) discharge from a similar culture at first in continuous darkness but transferred to light (1700 lx) at the indicated point. Time in days from inoculation. After Callaghan (1969).

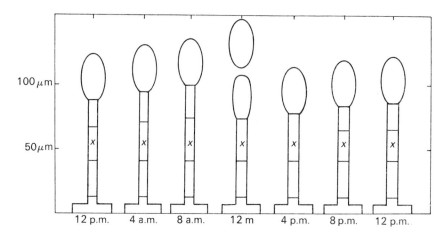

F_IG. 12.21. *Erysiphe polygoni.* Diagrammatic representation of the abstriction of conidia and of the changes in the conidiophore throughout the day, as observed on mildewed clover plants in the field at Madison, Wis., 4 September 1933. The generative cell (x) is the third cell from below. After Yarwood (1930).

other species, there does not appear to be any periodic ripening or release of conidia evoked by the alternation of light and darkness. The diurnal pattern for conidia of this fungus, with on the average a peak at noon, reported by Sreeramulu (1964), may be attributed to wind, which tends to a maximum around midday, rather than to light-induced periodicity.

In rhythms of spore liberation induced by light, essentially rhythms in *sporulation* are involved. Light rarely has a direct effect on take-off; an exception being the immediate puffing produced in *Ascobolus crenulatus* on illumination with blue light.

In general, circadian rhythms of sporulation are familiar to mycologists, if only as concentric zoning in cultures. Usually day-time zones of intense sporulation alternate with night-time ones where fewer spores are formed. Nearly always zonation can be related to the natural alternation of light and darkness (e.g. Hall 1933; Sagromsky 1952a, b), although temperature alternation may also induce zoning (Hafiz 1951), and occasionally the same effect is brought about as an endogenous rhythm without periodicity in external conditions being involved (Brandt 1953).

13. DISPERSAL BY INSECTS

IT is one of the most striking facts of botany that transport of pollen grains in angiosperms from flower to flower is largely the work of insects, and this reliance on insects has apparently been a major factor in the evolution of floral structures. On the other hand, although there are a number of interesting examples of spore dispersal by insects, fungi have tended largely to be anemophilous. Only in one major taxon, Phallales, have insects become the normal agents of dispersal and only in this order have they clearly affected the evolution of fungal form in a manner comparable with their impact on the evolutionary development of angiosperms. Phalloids have, however, already been discussed in the general context of Gasteromycetes.

On a smaller scale there are several types of fungi in which specialization in relation to insect dispersal seems to have occurred. A striking example is the conidial stage of *Claviceps purpurea*. This produces its sclerotia (ergots) in rye (*Secale cereale*) and in other members of Gramineae. Primary infection is due to filiform ascospores violently discharged during the late spring and borne by the wind to flowers of the host. The fungus invades the ovary, which becomes replaced by a dense mass of fungal tissue that is highly convoluted and covered by a hymenium of simple conidiophores. These produce vast numbers of very minute oval conidia that become involved in a copious, stinking 'honey-dew' in which the concentration of sugar may be over 2 molar. The flowers of grasses are not insect-pollinated, but some insects regularly visit the spikelets of rye for the sake of pollen. Others are attracted by the 'honey-dew' secreted by the leaf-louse, an insect found abundantly on the fertile spikes of rye. Insects, especially *Melanostoma mellina*, which feeds both on pollen and on 'honey-dew', whether of insect of fungal origin, seem to be normal vectors of the ergot disease. The spores may be carried on the surface of the insect or they may be sucked in with the spore-containing honey-dew, to be excreted or regurgitated later, perhaps on a healthy spikelet (Atanasoff 1920).

It should be mentioned that, although insects would appear to be the principal agents of the spread of the conidial (*Sphacelia*) stage of ergot, other methods of dispersal may sometimes obtain. In the unnatural conditions of field cultivation conidia may easily be spread by a diseased head knocking against a healthy neighbouring one, and probably rain-splash also plays some part in dispersal.

The importance of insects in the dispersal story of rust fungi, though envisaged by Ráthay as early as 1883, was brought into prominence by the brilliant work of Craigie (1931) who demonstrated that *Puccinia helianthi* and *P. graminis* are heterothallic (self-incompatible). This condition has since been shown to occur in many more rusts belonging to a wide range of genera (*Cronartium, Melampsora, Gymnosporangium, Phragmidium, Uromyces*). Indeed, self-incompatibility seems to be the normal state of affairs and, according to Buller (1950), it probably obtains in all rusts with fully developed pycnia. Homothallic (self-compatible) rusts seem few in number, and in them the pycnia are rudimentary or absent.

On germination of the teliospore of *Puccinia graminis* uninucleate basidiospores (sporidia) are produced. These are violenty discharged and, being further dispersed by wind, normally fall singly on the leaves of the host. If a basidiospore infects a leaf of barberry, a localized intercellular mycelium is produced on which pycnia are formed, mainly at the upper surface, and also spherical proto-aecia embedded in the somewhat hypertrophied leaf tissue near the lower surface. Each pycnium (Fig. 13.1) is more or less flask-shaped, with its wall of interwoven hyphae within the host tissue but with its ostiole, surrounded by a fringe of stiff red projecting periphyses, opening to the outside. Within the pycnium minute pycniospores are abstricted and ooze out through the ostiole together with nectar, so that each pycnium is capped by a little sugary drop having an osmotic pressure of between 12 and 24 atmospheres. Often at a later stage the drops from neighbouring pycnia flow together to form a single larger drop. In addition to the coloured periphyses there project into the nectar drop through the ostiole a few rather long 'flexuous hyphae' which are simple or sparingly branched and non-septate.

Apparently in most rusts there is a sweet scent associated with the pycnial stage. This may most easily be detected in the common thistle rust (*Puccinia suaveolens*) which causes a systemic infection of *Cirsium arvense*. Early in the growing season the under surfaces of the leaves of infected shoots are densely covered with pycnia. The diseased leaves have a scent, almost like that of violets, which disappears when the succeeding urediospore stage develops and the pycnia disappear. In such a rust as *Puccinia graminis*, where only small local infection spots are involved, the number of pycnia crowded together is very much less and the scent can be detected only by those with a keen sense of smell.

In Craigie's experiments with *P. graminis* and *P. helianthi* infection of a leaf by a basidiospore led to the development of an infected spot bearing pycnia at the upper surface and spherical proto-aecia near the lower surface. These did not normally develop into aecia if precautions were taken to exclude insect visitors. When, however, the pycnial nectar from one infection spot was added to the nectar around the ostioles in another single-spore infection spot, mature aecia developed in a few days, provided the two infections were of opposite mating type.

According to genetical evidence each aeciospore contains a 'plus' and a 'minus' nucleus. A proto-aecium borne on a 'plus' or a 'minus'

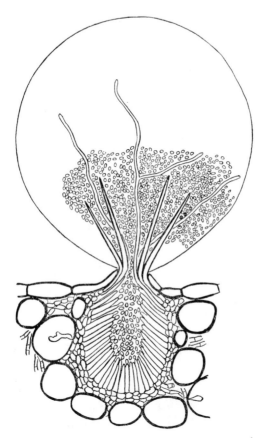

Fig. 13.1. *Puccinia graminis.* Single pycnium. The nectar drop contains numerous exuded pycniospores, and projecting into it are stiff, thick-walled, red periphyses and the thin-walled and sometimes branched flexuous hyphae. × about 300. After diagram by Buller (1950).

mycelium must receive nuclei of the opposite sign if it is to develop into an aecium. The pycniospores would seem to act as dikaryotizing agents although the cytological details of the process are still very obscure. When pycniospores are added to the drop around the ostiole of a pycnium of opposite sign, the introduced spores fuse with the flexuous hyphae in the drop. Presumably the nuclei of these spores pass into the flexuous hyphae, thence into the intercellular mycelium, and so to the proto-aecium. Here dikaryotization of certain cells occurs and it is these that give the chains of aeciospores.

Craigie has shown that under experimental conditions flies will mix nectar from different monosporidial cultures and so lead to the development of mature aecia. Very much earlier Ráthay (1883) studied the insect visitors to pycnial nectar and recorded a large number of insects that were attracted, including flies, beetles, and ants.

An example somewhat similar to rusts is *Coprinus cinereus* (Brodie 1931). This coprophilous agaric has the tetra-polar self-incompatibility type of heterothallism that is a feature of most of the larger Basidiomycetes. Each basidiospore gives rise on germination to a monkaryotic mycelium of uninucleate cells, and from this conidia (oidia) are produced on short conidiophores in slimy heads. The monokaryotic mycelium can develop into a dikaryotic one bearing fruit-bodies only if it receives nuclei of the appropriate mating type, and one of the ways this can happen is by contact with suitable oidia. It has been shown experimentally in cultures that flies may transport oidia and effect dikaryotization of a monokaryotic mycelium. No doubt this kind of transport is also effective in nature, although it must frequently, perhaps usually, occur following the intermingling of monokaryotic mycelia derived from basidiospores of compatible mating types germinating in the same dung ball.

In some pathogenic fungi spore production is in the host flower and dispersal from infected to healthy plants is brought about by insects in the course of pollination. The most familiar example is *Ustilago violacea*, the common anther-smut of campions and other members of Caryophyllaceae. Infection of the host is systemic or semi-systemic but spore production is restricted to the anthers. Diseased plants appear almost normal until flowering, when it is found that blackish-purple smut spores are produced in place of pollen grains. In *Lychnis alba* there are separate staminate and pistillate plants. When the staminate plant is infected by *U. violacea* the floral structure is modified only to the extent that smut spores are formed in place of pollen. When, however, the pistillate plant is infected the staminal rudiments are stimulated to develop into stamens, but from their anthers brand spores, not pollen grains, are set free (Fig. 13.2). Further, the ovary is somewhat reduced and although ovules are present they do not develop into seeds. The smutted flowers still give out their scent and are visited by night-flying moths that distribute the spores to flowers of healthy plants. There is conflicting evidence about how infection of the host takes place (Baker 1947), but it seems clear that when brand spores are transferred to a flower on a healthy plant they germinate to produce a mycelium that grows back through the flower stalk into the shoot, so that flowers that develop subsequently on that plant may be smutted. The old view that contamination

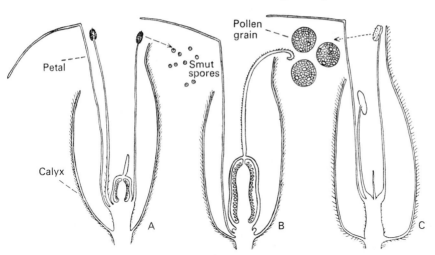

FIG. 13.2. Flowers of *Lychis alba* seen in longitudinal section. (A) Infected pistillate flower showing stamens (developed from staminal rudiments) containing smut spores. (B) Pistillate flower from healthy plant showing staminal rudiments. (C) Staminal flower from healthy plant. Smut spores and pollen grains drawn to same scale and greatly magnified.

of a flower by smut spores led to the production of infected but viable seed, as in the loose smut of wheat and barley (*Ustilago nuda*), is not borne out by experiment. Again, infection through the flower is probably not the only method by which the fungus is handed on to healthy plants. There is little doubt that seedling infection also occurs.

Another anther-smut in which the spores are probably dispersed by insects is *Ustilago succisae*, found frequently on devil's bit scabious (*Succisa pratensis*). Here spore production is again limited to anthers. However, the brand spores are white or pale cream in the mass, not black, as in most smuts. Infected flower-heads are easily spotted since the projecting anthers are whitish in contrast to their purple colour in normal plants.

The anther mould of red clover caused by the hyphomycete *Botrytis anthophila* (Silow 1933), shows a remarkable parallelism with the anther smut of campions. The disease is systemic, but spore production is limited to the flowers, the conidiophores of the fungus being produced on the surface of the anthers (Fig. 13.3). The conidia may be carried by a pollinating bee to the stigma of a healthy flower, where they germinate, and the germ-tubes penetrate the tissue of the style after the fashion of pollen-tubes The developing seeds are infected and within these a dormant mycelium is formed. These infected seeds are viable, but in the following year they give diseased plants.

A further example of dispersal of a fungal pathogen by pollinating insects is to be found in endosepsis of figs (Caldis 1927). The causal

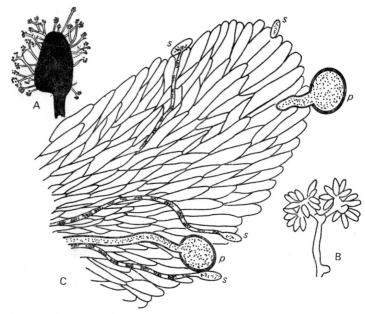

FIG. 13.3. *Botrytis anthophila.* (A) anther and part of filament from the flower of a diseased clover plant; the anther is covered with the conidiophores of the fungus; (B) a single conidiophore bearing numerous conidia; (C) the stigma of a clover flower showing germinating pollen grains (*p*) and conidia (*s*). After Silow (1933).

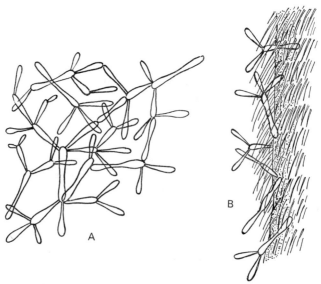

FIG. 13.4. *Candida reukaufii.* (A) colony of cells; (B) groups of cells attached to the proboscis of a bee. Highly magnified. After Grusz (1927).

organism (*Fusarium moniliforme* var. *fici*) may be introduced into the edible fig by its obligate pollinator, the fig-wasp (*Blastophage psenes*). Again, nectar-fermenting yeasts are often carried from flower to flower by bees. For example, *Candida reukaufii* (Grüsz 1927) grows in the nectar of a wide range of flowers. The fungus can be isolated by trapping a bee after it has visited an affected flower, rendering it unconscious with ether, and finally cutting off its proboscis and planting it on nutrient agar (Fig. 13.4). It is thought that the yeast may spend the winter in the alimentary tract of bumble bees and be introduced into nectaries in the spring (Hautmann 1924).

A number of fungi fructify in the bore-holes and brood galleries of wood- and bark-beetles, which not only disperse the spores but also introduce the fungi into the inner tissues of the host. The best-known example is Dutch elm disease, which has killed so many elms in the past fifty years. In this disease bark-beetles (chiefly *Scolytus* spp.) are responsible for the transport of the spores of the pathogen, *Ceratocystis ulmi*. The insects form characteristic 'engravings' (brood galleries) in the dead or dying elm trunks at the interface of wood and bark. The eggs are laid in the main gallery and the insects develop from egg to pupa in the lateral galleries, on the walls of which the fungus produces its heads of sticky conidia (Fig. 13.5).† Finally, the young beetles bore their way directly outwards through the bark from the widened ends of the lateral channels, leaving little round holes in the bark so that from the outside it seems to have been peppered by shot. The emerging insects carry with them, both internally and externally, viable spores of the fungus. *Scolytus* can produce its brood galleries only in the larger limbs and trunks of dead and dying trees, but before doing so it feeds on the bark in the crutches of very small branches of healthy trees. This introduces the pathogenic fungus to the wood of vigorous living trees which, as they succumb in the course of years, become suitable sites for brood galleries.

The fungus *Ceratocystis ips*, which causes a 'blue stain' in the wood of conifers, has a similar dispersal story (Leach 1940), being spread by bark-beetles (*Ips* spp.). However, only trees already dead or very much weakened are affected because the bettles do not, as in Dutch elm disease, prepare their future victims by feeding on the younger healthy parts before boring into the dead trunks for breeding purposes. However, the blue-stain fungus, though so closely similar to the elm parasite, occurs mainly in the perithecial stage, but this is biologically just like the conidial stage of *Ceratocystis ulmi*, for a mass of slimy ascospores is

---

† The perithecial stage seems rarely to be produced in nature, but from the point of view of spore dispersal it is like the conidial stage, an extruded drop of slimy ascospores produced at the end of a long perithecial neck.

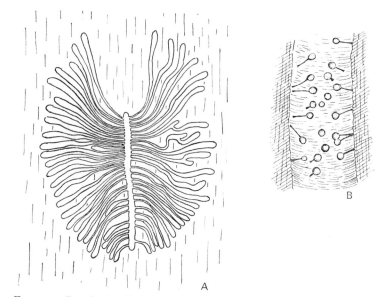

Fig. 13.5. Dutch elm disease. (A) brood galleries of bark-beetle (*Scolytus* sp.) at interface of wood and bark; (B) small part of gallery magnified, showing the white heads of slimy conidia of *Ceratocystis ulmi* borne on black stalks. (A) after McKenzie and Becker. (B) based on photographs by Clinton and McCormack.

borne at the end of a little stalk, but the stalk is the long neck of a perithecium the base of which is buried in the wood of the brood gallery.

The story of dispersal in the fungi (especially *Leptographium* spp.) 'cultivated' by the ambrosia beetles is somewhat similar (Leach 1940, Webb 1945, Bakshi 1950). The beetles bore deeply into the wood of dead and weakened trees, producing branching galleries. The fungi grow in the wood immediately surrounding the galleries and cause a characteristic darkening of the tissue. Sporulation occurs on the gallery walls and the spores are used by the grubs as food. When beetles migrate to another tree, they carry viable spores of the fungus with them and so a continual supply of 'ambrosia' is assured.

A somewhat similar association of insect and fungus is that of the fungus-growing ants (Uphof 1942). These ants cultivate certain fungi in their nests in a state of more or less pure culture. When a queen establishes a new colony she carries an inoculum of the fungus with her in her infrabucal pouch.

Again, wood wasps may enter into special relations with certain lignicolous fungi. An interesting example is the great wood wasp, *Sirex gigas*, which attacks coniferous trunks. The female has a long ovipositor (Cartwright 1926, Parkin 1942). Near the fixed end of this in the abdomen of the insect are two little pouches in which *Stereum sanguinolentum* is to be found in the oidal condition. As an egg enters the ovipositor from the main body of the insect an oidial inoculum is exuded on to it from the pouches so that as each egg is laid in a coniferous trunk the fungus is introduced at the same time. The advantage to the insect of this regular association is far from clear. Perhaps the fungus pre-digests the wood for the larva, or perhaps the larva is, in fact, mycophagous. So far as the fungus is concerned this means of dispersal, though very efficient, is merely additional to its regular spread by air-borne basidiospores.

In connection with insect dispersal of fungi growing on tree trunks, *Cryptoporus volvatus* (Fig. 13.6) is of considerable interest (Hubbard, 1892, Buller n.d.). It is a smallish bracket polypore, found in North America and northern Asia on dead and dying trunks of conifers severely attacked by bark-boring beetles. When a fruit-body is about to be formed the mycelium accumulates in an old abandoned bore-hole and where this emerges the sporophore is formed. Both in general appearance and in texture it is rather like a small specimen of *Polyporus betulinus*, but the tube layer is covered in below by a firm volva. At first this is continuous, but later a small round hole is formed in it. Spores rain down from the hymenial tubes as in any other polypore, and a very small proportion of

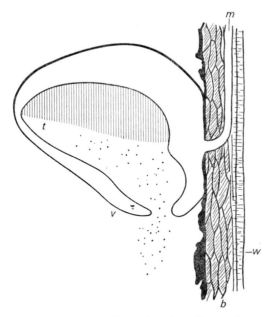

FIG. 13.6. *Cryptoporus volvatus*. Vertical section of sporophore growing on a dead conifer. The sporophore is formed at the end of a mycelial strand (*m*) filling a boring made by a bark-beetle between wood (*w*) and bark (*b*). Some of the spores falling from the hymenial tubes (*t*) are seen escaping through the hole in the volva (*v*). Drawn from a half-tone figure by Henessey in Buller (n.d.).

these fall through the hole and thus escape into the outer air. The vast majority, however, are deposited on the inner surface of the volva. Through the hole in the volva, no doubt attracted by the dimness within, many insects crawl. These, mainly beetles though not bark-beetles, become contaminated with numerous spores and on emerging may go on to explore abandoned bore-holes made by bark-beetles in the trunk of another tree. So, perhaps, the fungus is dispersed, though no doubt the spores that escape directly through the hole in the volva may infect nearby trunks. The dispersal story is in need of closer study, but in view of the unique structure of the fruit-body the fungus is more likely to be ento-mophilous than anemophilous.

Hemipterous insects with their piercing sucking mouth-parts are of great importance in the transmission of virus diseases of plants. They may also pick up the spores of certain pathogenic fungi during feeding and later introduce them into healthy plants. This type of disease is found particularly in the fruits of a wide range of species and is referred to as stigmatomycosis. The fruits are affected internally but remain sound on the outside. An example of considerable economic importance that has been much studied is the internal boll disease of cotton (Fraser 1944). This is caused by *Ashbya gossypii* and allied fungi and leads to staining of the lint in immature, unopened bolls. The fungus grows and sporulates in the lint, but the pathogen makes no appearance at the surface of the boll. The chief vectors of the disease are bugs of the genus *Dysdercus*. When a bug feeds on a diseased boll the tip of its fine sucking apparatus punctures the ovary wall. This apparatus, throughout the greater part

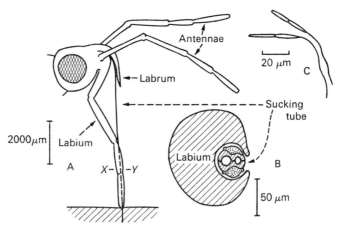

Fig. 13.7. (A) head and appendages of plant bug *Dysdercus*, showing piercing-sucking apparatus puncturing the skin of a fruit; (B) T.S. of labrum in region *X–Y* of A, showing the labium as a split tube, and the sucking tube consisting of four mouth-parts (two mandibles and two maxillae) closely set together and thus forming two tubes; (C) spores of *Ashbya gossypii*. A, B based on figures by Fraser (1944). C after Ashby and Nowell (1926).

of its length, is sheathed in the split tube formed by the labium, the tip of which rests on the surface of the fruit. The sucking apparatus is composed of four stylets (actually a pair of maxillae and a pair of mandibles) tightly pressed together, and contains two longitudinal tubes: a narrower one down which saliva is pumped and a slightly wider one up which the food solution is sucked (Fig. 13.7). The long, narrow ascospores, each with a whip-like tail, are sucked up with the liquid food and may be stored unharmed in the stylet pouches that occur in the head. Others may pass down the alimentary canal but these, apparently, soon cease to be viable. When a contaminated insect proceeds to a young and healthy boll, it may introduce the pathogen. Just how the spores pass from the stylet pouches into the host plant is not clear, but perhaps they find their way into the top of the salivary canal and are pumped thence into the host.

In a rather similar but much more casual manner, wasps may spread *Monilia fructigena* amongst ripe fruit during the late summer. This 'brown-rot' organism is abundant on apples, pears, and plums, both on fruit attached to the tree and on windfalls. The rot rapidly spreads throughout an infected fruit and later, on its surface, pustules of powdery spores are formed in characteristic concentric rings. Entry is possible only through a wound and this is often provided by the bite of a wasp, which, if it has already visited a rotten fruit bearing conidia, may introduce the pathogen at the same time. This is probably the normal method by which the fungus spreads, but the spores are dry and powdery and no doubt wind dispersal occurs to some extent.

This discussion of entomophilous dispersal is by no means exhaustive and many examples have been omitted. Two points, however, stand out clearly: first, the types of insect concerned are very various (flies, beetles, moths, plant bugs, ants, bees, and wasps); second, the fungi involved are equally diverse (stink-horns, a bracket polypore, flask-fungi, smuts, rusts, yeasts, and imperfect fungi). No doubt with further work instances of insect dispersal will be multiplied, but the general statement that entomophily is not common in the fungi is likely to remain unassailed.

It should be observed that other small invertebrate animals may be responsible for the spread of fungi. Slugs, for example, frequently feed on the larger agarics and, apparently, the spores pass uninjured through the alimentary canal and then germinate freely. Indeed, it has been claimed that slugs play a very important part in the dispersal of fleshy toadstools, but this seems unlikely (Voglino 1895). Again, mites may be effective agents in the spread of moulds. Sometimes in a mycological laboratory they become established in a few tube-cultures of moulds and then with alarming speed contaminants appear in nearby pure cultures. Even eel-worms may disperse fungal spores. Thus in the *Dilophospora* disease of cereals Atanasoff (1925) claimed that it is nematodes that normally carry spores to the susceptible growing-point of the shoot.

14. DISPERSAL BY LARGER ANIMALS

LARGER animals play a part in the dispersal of certain fungi. Thus the coprophilous species are dispersed by herbivores which also provide, in the form of their droppings, an ideal culture medium. Again some wood-land hypogeal fungi, the truffles and their like, are apparently dispersed by rodents. Further, waterfowl almost certainly play an essential part in the long-distance spread of freshwater aquatic fungi, a matter that will be discussed in the final chapter. Moreover, it is logical to refer here to the activities of man who has, unwittingly, been responsible for the spread of many fungal pathogens of plants even from one continent to another.

If the freshly deposited dung of herbivorous animals (e.g. horse, sheep, rabbit) is kept reasonably moist and adequately aerated under a bell-jar in the light, a rich and characteristic fungal flora develops. There is a fairly definite succession with mucoraceous fungi (especially species of *Mucor*, *Pilaira*, and *Pilobolus*) in the van, followed by Ascomycetes, species of discomycete genera (e.g. *Ascobolus*, *Ascophanus*, *Rhyparobius*) tending to appear ahead of pyrenomycete genera (e.g. *Podospora*, *Sordaria*, *Sporormia*, *Chaetomium*). Last of all a crop of agarics is produced usually consisting of small species of *Coprinus*. The stages often overlap and the succession under field conditions may not be so regular as in the labora-tory, but the general story of succession is usually fairly definite.

The causes of this ecological succession are not yet fully understood but the work of Harper and Webster (1964) helps to resolve some of the problems. A basic factor determining the successive appearances of the different fungi seems to be the time interval between germination and the production of spore-bearing structures. On sterile pellets of rabbit dung inoculated with single species, this interval is 2–4 days for the mucora-ceous fungi but 1–2 weeks for *Coprinus* spp.

The general story of dispersal in most coprophilous fungi seems to be the same. The spores produced by the sporophores on the dung reach the surrounding grass which may later be eaten by a horse, sheep, or rabbit. The spores not only pass uninjured through the alimentary canal of the animal, but are exposed there to conditions stimulating germination which takes place in the deposited dung. Spores of some coprophilous species germinate only if treated to some of the conditions encountered in the intestine, such as relatively high temperature or an alkaline reaction or both. However, some can germinate without any such treatment.

Spores of *Coprinus* spp. occurring on dung need no special conditions, and, although the germination of *Pilobolus* spores is greatly stimulated by a somewhat high temperature, a small percentage do germinate under more ordinary conditions. Since the spores of some dung fungi can germinate without passage through an animal, some of the fungi that make their appearance on dung may develop from airborne or from insect-borne spores, but there is little doubt that for the coprophilous flora as a whole the normal channel of dispersal is through the animal.

So far as their apparatus of dispersal is concerned some dung fungi seem little, if at all, modified in connection with their particular habitat. No obvious specialization occurs in the coprophilous species of *Mucor*, *Ascophanus*, *Chaetomium*, or *Coprinus*. However, fungi such as *Pilobolus*, *Dasyobolus immersus*, and *Podospora* spp. show, in the organization of their spore discharge, what seem to be beautiful and parallel adaptations. This apparent specialization is associated with the first stage of dispersal, namely the passage of spores from the dung to the grass. In these fungi the spores are discharged to such a distance (10–200 cm) that they can reach the grass without the aid of wind. We have seen that the distance D to which a microscopic spherical projectile (of radius r) is shot when discharged with a given initial velocity approximates to $D = Kr^2$. The long range of the spore-guns in these specialized coprophilous fungi is due rather to the comparatively large size of the projectile than to exceptionally high initial velocity at take-off. In some coprophilous fungi the individual spores are small, but are bound together by mucilage to form large projectiles. This is so in *Pilobolus* spp. in which the whole sporangium, containing hundreds or thousands of spores, is the projectile; in the minute discomycete *Rhyparobius nanus*, where the ascus contents of 200–300 spores are cemented together by slime; and in the pyrenomycete *Podospora setosa* in which the 128 spores, individually small, form a bulky projectile. In some other species, notably *Dasyobolus immersus* and *Podospora fimicola*, the eight spores of the ascus are not only stuck together by mucilage, but also are relatively enormous (Fig. 14.1).

In some of the more specialized fungi growing on the dung of herbivores the apparent parallel development in spore-discharge equipment in striking (Fig. 14.2). This may be emphasized by considering three common species: *Pilobolus kleinii*, *Dasyobolus immersus*, and *Podospora fimicola*. In all the spore-projectile is relatively large, the spores being held together by mucilage. In all the distance of discharge is considerable; 2–3 m in *Pilobolus* (Buller 1934), and about 30 cm in *Dasyobolus immersus* (Buller 1909) and in *Podospora fimicola* (Ingold 1933). Thus the spores can be thrown onto the surrounding grass without the aid of wind. In all discharge is by day and, further, the spore-gun is orientated by the positive phototropism of the sporangiophore in *Pilobolus*, of the individual projecting asci in *Dasyobolus*, and of the perithecial neck in *Podospora*.

This phototropism helps to ensure that the spore-projectiles are thrown clear of the substratum. In all three fungi the projectile is strongly adhesive, the mucilage on drying cementing it firmly to the surrounding stems and leaves. There is evidence that strong light has an injurious effect on fungal spores, and it is therefore of interest to note in these fungi that the discharged spore-mass on the grass has the protoplasm shaded

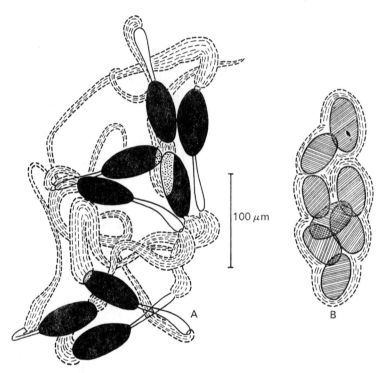

100 μm

FIG. 14.1. Group of ascospores discharged from an ascus of *Podospora fimicola* (A) and from an ascus of *Dasyobolus immersus* (B).

by the pigment located in the spore-wall of *Dasyobolus* and *Podospora* and by the black part of the sporangial wall in *Pilobolus* which effectively shields the colourless spores (see p. 76).

A strong case can be made (Buller 1934) for adding *Sphaerobolus stellatus* to the list of specialized coprophilous fungi. It is quite common on old cow-dung pads although it is, perhaps, more familiar as a lignicolous species. When coprophilous there is no doubt that its dispersal story is essentially like that of *Pilobolus*. It has all the features of the specialized fungi of dung with its relatively large, adhesive glebal-mass projected to a distance of several metres and aimed towards the light by the phototropic response of the sporophore. Further discharge is by day and the spores and gemmae of the glebal-mass are shaded from strong light by a thin dark-brown membrane.

It must again be emphasized that by no means all the dung fungi exhibit specialization in relation to this habitat. It is especially true of the small coprophilous agarics, but it is interesting to notice that these tend to have black or dark-brown spores, so that again the contained protoplasm

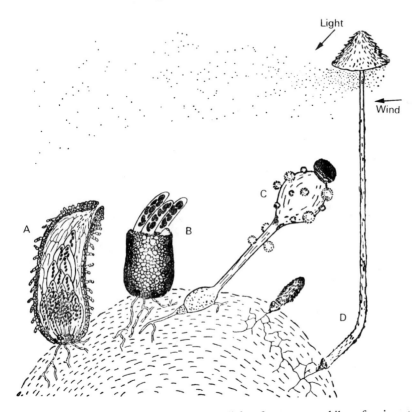

FIG. 14.2. Diagram showing reaction to light of some coprophilous fungi. (A) *Podospora tetraspora* (seen in optical section), × 50. (B) *Dasyobolus immersus*, × 40. (C) *Pilobolus kleinii*, × 10. (D) *Coprinus fimetarius: left:* young sporophore primordium orientated by positive phototropism; *right:* mature expanded fruit-body liberating spores, nat. size.

is protected from light during the sojourn of the spores on the grass before it is eaten. Although dark-spored agarics are abundant in other habitats, the absence of species with colourless spores does seem significant.

Just how the spores of the coprophilous Mucorales (especially *Mucor* spp. and *Pilaria anomala*) reach the grass from the dung is not clear. It has already been noted that most species of *Mucor* form sporangial drops. A projectile of *Pilobolus* might, immediately after discharge, brush against such a drop, picking up some spores that might then be carried as passengers to the surrounding herbage. That this does happen can easily be demonstrated by catching *Pilobolus* sporangia in mid-air on the surface of sterile nutrient agar, when a growth of *Mucor* frequently develops

Fig. 14.3. Mycologists raking for truffles during a fungus foray at Hereford. Reproduced from a small panel of a full-page illustration by Worthington Smith in the *Graphic* of 1873.

from the projectile, but this seems improbable as the main method of primary dispersal; perhaps splashing rain or insects play a part.

On the excreta of frogs a minute phycomycete, *Basidiobolus ranarum*, occurs with great regularity (Levisohn 1927). It seems to be on its own and no coprophilous flora develops as on the dung of herbivores. Interest centres on the fact that the fungus is so like a tiny *Pilobolus* although, systematically, there is a considerable gulf between them. How the conidium is discharged to a distance of several centimetres has already been discussed. But this is not the end of the story of dispersal. Small beetles feed on the discharged spores, but before the spores have lost their viability the insects may be devoured by a frog. In its intestine the conidium, which is really an immature sporangium, completes its development and gives rise to eight sporangiospores. These are voided with the excreta and germinate to give mycelia there.

The hypogeal fungi form an interesting biological group. The fruit-bodies, varying from a few millimetres to several centimetres in diameter, are buried in the soil beneath the leaf carpet in woods. In most of them the story of dispersal would seem to be the same, but observations are few and experimental work is apparently completely lacking. In a number of these fungi the fruit-body, when mature, gives off a smell that attracts rodents. These grub up the fruit-bodies and eat them. The spores no doubt pass through the alimentary canal and are distributed in the droppings. But these fungi are not coprophilous. They are soil fungi and some, perhaps all, enter into mycorrhizal relations with forest trees.

In hypogeal fungi convergent evolution has clearly occurred, the same type of fruit-body having evolved in Endogonaceae (a family of Mucorales), in Hymenogastraceae (a family of Basidiomycetes) and in two very distinct families of Ascomycetes (Elaphomycetaceae and Tuberaceae). Even in Tuberaceae there does not appear to have been a single line of evolution.

Perhaps the commonest British hypogeal fungus is *Elaphomyces granulatus* (hart's truffle), occurring both in deciduous and coniferous woods. Often the little excavations are to be seen where rodents have grubbed for the fruit-bodies and sometimes an abandoned half-nibbled specimen may be discovered left behind, no doubt, when the animal was disturbed. It is well known that the French, much more inclined to mycophagy than the British, train dogs and pigs to locate truffles, but which animals are mainly responsible for dispersal in nature does not seem to have been studied.

Hypogeal fungi are usually regarded as rather rare, but this is only because few have taken the trouble to look for them. In Britain many records were made last century largely due to the activities of Broome who, armed with a hand-rake, searched methodically for these fungi (Fig. 14.3). Sustained raking in forests, combined with sharp eyes, will almost invariably lead to the unearthing of these hypogeal fungi.

There is probably a certain amount of casual dispersal of resistant fungal spores that are able to pass uninjured through the alimentary canal of larger animals. Sometimes dispersal of this kind may be of economic importance. An example is *Plasmodiophora brassicae*, causing club-root of crucifers. If diseased turnips are fed to cattle, the minute resting spores get into the dung. If this is used as manure viable spores are added to the soil and may cause infection in a cruciferous crop. For this reason some gardeners prefer to use artificial fertilizers rather than run the risk of introducing this fungus by the use of natural manure.

Birds have occasionally been suspected of transporting fungal spores and there is evidence that they may have played a part in the disastrous spread of the bark disease of American chestnut in the second decade of this century. Plant pathologists in the United States made a thorough study of the dispersal of the casual organism *Endothia parasitica*. The local spread of the disease from tree to tree in a single wood was clearly due to airborne ascospores discharged from perithecia during wet weather, but it was thought that the occasional long-distance dispersal to remote plantations might be due to ascospores or pycniospores carried by birds, especially woodpeckers and tree creepers, on their breasts, tail feathers, feet, and beaks (Heald and Studhalter 1914). Birds were shot in the neighbourhood of infected trees and a 'poured plate' technique was used in estimating the spore load on each bird. Two downy woodpeckers were each found to be carrying over half a million and a brown tree creeper over a quarter of a million viable spores. The general conclusion was reached that many of the local centres of infection, isolated from the general area of chestnut disease, might well have originated from spores carried by these birds.

It would be long and rather tedious to discuss in detail the influence of man on the dispersal of fungi. Occasionally he deliberately aids in dispersal as when he distributes mushroom 'spawn' or cultures of moulds for the manufacture of cheese, but usually dispersal is an incidental concomitant of his global activities such as air travel (Baker 1966).

It is in the intercontinental spread of pathogenic fungi that man's influence has been most marked. Such vast expanses as the Atlantic and Pacific Oceans appear to offer formidable or perhaps insuperable barriers to the spread of fungal spores. But where nature has failed man has unwittingly succeeded. The story of plant pathology in Europe is punctuated by invasions from the Western Hemisphere, for example potato blight (introduced about 1840) and American gooseberry mildew (about 1900). And the North American continent has received its share of pathogens from Europe such as the blister rust (*Cronartium ribicola*) of white pine (introduced about 1906) and Dutch elm disease (about 1930). Further, at least one serious disease in North America, the bark disease of chestnut, appears to have originated in China or Japan.

The spread of fungal diseases from one continent to another has, apparently, been due largely to the introduction of infected material (e.g. nursery stock) of the host plant rather than to the accidental transport of individual fungal spores. Through bitter experience man has now learnt that great care must be taken to prevent the spread of disease-producing fungi to new areas, and most countries have taken steps, often fairly effectively, to guard against the introduction of further pathogenic species from abroad (Stakman 1947).

15. SEED-BORNE FUNGI

A NUMBER of fungi are dispersed in an association of some kind with the dispersal units of the host. These units are usually seeds, or seed-like fruits such as the caryopses of cereals or the mericarps of umbellifers. Although seed-borne fungi have but a small place in the general picture of fungal dispersal, they are of great significance to the plant pathologist since many important plant diseases are seed-borne. Indeed, there are few plants of major economic importance which have not at least one seed-borne fungal pathogen (Malone and Muskett 1964).

There are a number of endophytic fungi, causing systemic infections of their hosts, that are normally seed-dispersed and, at least under natural conditions, do not seem to be dispersed in any other way. Thus in members of Cistaceae an endophytic fungus is regularly present. The story has been studied in some detail in *Helianthemum chamaecistus* (Boursnell 1950). The fungus is found in root and shoot, and when flowers are produced enters the ovary wall. The outer gelatinous coat of the testa is infected, but the fungus is prevented by a barrier of tannin-filled cells from passing into the embryo and endosperm. Infection of the seedling from the dormant mycelium in the testa occurs at germination and under natural conditions such infection seems to be necessary for continued growth.

Another fungus that apparently relies entirely on seed dispersal is the well-known endophyte of *Lolium* spp. (Freeman 1904). The fungus forms an intercellular mycelium throughout the shoot system of the host. In due course the ovary is infected and the caryopsis carries a dormant mycelium in the testa (Fig. 15.1). With some difficulty the fungus has been isolated in pure culture, but no spores were produced so, as with the endophyte of *Helianthemum*, its systematic position is unknown. From *Lolium* a second endophyte can sometimes be isolated which grows well in culture and produces microconidia. Although most specimens of *Lolium* are infected some are free from either endophyte. On making reciprocal crosses between infected and uninfected plants it has been shown that the fungus is 'inherited' only through the ovule (Sampson 1935, 1937, 1939).

Another interesting fungus that is sometimes seed-borne is *Epichloe typhina*, the 'choke' of grasses. This attacks a number of species belonging to several genera causing systemic infection. In *Dactylis glomerata*

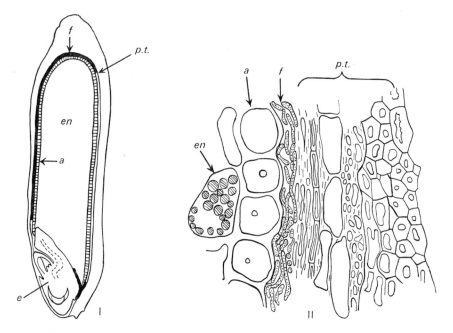

FIG. 15.1. (I) L.S. grain of *Lolium*, and (II) small part much more highly magnified. (e) embryo; (en) starchy endosperm; (a) aleurone layer; (p.t.) fused pericarp and testa; (f) layer of hyphae of endophytic fungus. After Freeman (1904).

flowering is prevented in affected plants and the disease cannot, therefore, be seed-borne. However, when *Festuca rubra* is attacked (Sampson 1933) it still produces seed and these may carry the fungus as a resting mycelium in the embryo and endosperm. Further, plants of *F. rubra* may show no external manifestations of disease, no stroma being formed. Where this type of latent infection occurs the fungus is dispersed solely with the seed of the host and a condition of affairs very like that of *Lolium* and its endophyte is attained.

Among important plant diseases that are normally seed-borne, the smuts of cereals are outstanding. Most smuts are systemic with the fine intercellular mycelium extending throughout the shoot system of the host plant. Spore production, however, is usually localized in and about the reproductive organs. In cereal smuts the ovary is the chief centre of spore production and no good grains are formed by infected plants, but instead of the grain there is formed a covered or uncovered mass of blackish brand spores. In stinking smut (bunt) of wheat caused by *Tilletia caries*, the wall of the caryopsis is unaffected but the interior is filled with black spores smelling of decaying fish due to the production of trimethylamine. During harvesting these grains are broken and the brand spores get dusted over healthy grains where they remain dormant until sowing time. The grains of wheat and the brand spores germinate at the same time and so infection of the seedlings is assured. Here the fungus is carried merely on the surface, so that the disease is easily controlled by shaking the grain in a drum with an organo-mercurial dust. It is of interest that, although this fungus is normally seed-borne, it is not necessarily so. Thus where an infected crop is threshed on the spot, spores may settle on the nearby soil, remain dormant but viable, and infect clean seed sown in the following season.

In *Ustilago nuda*, the loose smut of wheat and barley, the story is rather different. Spikelets of diseased plants are reduced to an exposed mass of brownish-black brand spores at a time when the uninfected plants are in flower. The dry powdery brand spores are wind-borne and may reach the stigmas of healthy plants, but instead of being resting-spores they germinate at once to produce a mycelium that penetrates the ovary and enters the developing seed. The seeds, however, are not destroyed but simply contain the fungus in a resting condition so that at germination the seedlings are already infected. In such smuts surface sterilization of the grain is useless and control depends on the fact that the fungus is more sensitive to heat than the embryo of the seed, so that steeping in hot (50–52°C) water for a short time (about 15 min) kills the parasite without greatly affecting the viability of the grain.

Another group of cereal diseases that are typically seed-borne are the 'stripe diseases' caused by *Helminthosporium* spp. *H. avenae* is responsible for leaf stripe and seedling blight of oats (Turner and Millard 1931). Except for a very local spread of the disease due to airborne conidia during the growth of the crop, dispersal is dependent on infected grains that have a resting mycelium on their surface and may also carry viable conidia, although their importance is relatively slight. *H. gramineum*, causing leaf stripe of barley, is similar, but the resting mycelium is usually in the grain between pericarp and testa.

Seed-borne fungi are specially prominent in the Gramineae, but they are also common in other families of flowering plants. Some of the diseases of pulses are seed-borne. A well-known example is anthracnose of French and runner beans due to *Colletotrichum lindemuthianum*. On all parts of the shoot this fungus produces small circular lesions in association with which slimy conidia are formed and are dispersed locally by rain-splash. Where the pods are infected the parasite may spread to the developing seed which ripen with the fungus established in the tissue of the cotyledons (Fig. 15.2). The appearance of disease in the following season is due almost entirely to the use of infected seed. The fungus is too far embedded in the embryo for seed sterilization to be effective, and control consists largely in the use of clean seed. The story of leaf, stem, and pod spot of peas caused by *Ascochyta pisi* is very similar.

Flax is of special interest because its principal fungal diseases are seed-borne (Muskett and Colhoun 1947). Thus seedling blight caused by *Colletotrichum linicola* is carried over from one season to the next as a resting mycelium in the outer layers of the seed-coat. Stem break and browning due to *Polyspora lini* is transmitted in much the same manner, and this is also true of the grey mould (*Botrytis cinerea*) which may

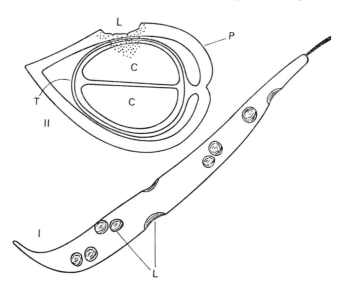

FIG. 15.2. Runner bean attacked by *Colletotrichum lindemuthianum*. (I) Diseased pod. (II) T.S. diseased pod and contained seed with region invaded by fungal hyphae indicated by dotting. (L) lesions caused by the parasite; (P) fruit wall; (T) testa of seed; (C) cotyledons of embryo. After Whetzel.

hibernate as a resting mycelium in the outer layers of the testa. However, *B. cinerea* is a very unspecialized and common parasite occurring on a large range of hosts and also as a saprophyte on dead plant material, but the early infection of the flax seedlings is commonly from the seed-borne mycelium, and only in connection with later infections are other sources of inoculum important. Other flax diseases such as foot rot, due to *Phoma* spp., and wilt, caused by *Fusarium lini*, may also be seed-borne, but infection of the new crop is often directly from the soil. In Northern Ireland, formerly a great flax-growing centre, the examination of flax seed for the presence of seed-borne fungi became an important aspect of seed-testing. In the method used (Ulster method) a random selection of seeds from a sample was placed on sterile malt agar in Petri dishes. These were incubated at 22°C and examined 5 days later. If a seed carried one of the important flax pathogens capable of rapid growth on agar (e.g. *Colletotrichum linicola, Polyspora lini, Phoma* sp., *Botrytis cinerea*, and *Fusarium lini*), a characteristic colony developed on the medium around the seedling. Although the seeds often carried surface spores of saprophytic moulds (especially *Penicillium* spp.), their growth did not, in practice, interfere with this simple and effective test, which could be carried out on a large scale and gave a clear picture of the degree to which the seed was contaminated and of the nature of the pathogens present.

So far examples have been considered where the fungus is seed-borne as a resting mycelium or as spores or both, but sometimes a whole fruiting structure is involved. This is so in *Septoria apii*, which causes leaf-spot of celery (Fig. 15.3). All parts of the shoot may be affected,

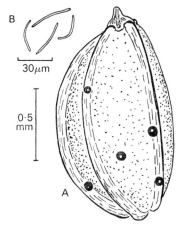

FIG. 15.3. (A) celery 'seed' (mericarp) infected with *Septoria apii*. Five pycnidia showing. (B) spores liberated from the pycnidium on moistening the 'seed'.

causing local infection-spots bearing pycnidia. From these, long pycnidio-spores ooze out embedded in slime and may be splash-dispersed. The developing fruits are often infected and on the mature mericarps ('seeds') pycnidia are frequently formed. These remain alive but dormant when the 'seeds' are dry, but when they are soaked and germination occurs, spores ooze out from the pycnidia and infect the seedlings.

Apart from those diseases in which the fungus is actually carried in some way or other on or in the viable seed, there are a number of fungal diseases where the pathogen, though not actually attached to the living seed, tends to be dispersed as a contaminant of the seed sample. This type of dispersal is somewhat unnatural, but may be of considerable economic importance. For example, a sample of rye seed may contain sclerotia of ergot (*Claviceps purpurea*), and at sowing-time these are scattered with the grain. Again, in the blind-seed disease of *Lolium* spp. (Wilson, Noble, and Gray 1945) caused by *Phialea temulenta*, infected seeds are not viable, but in a commercial sample may be mixed with healthy ones. Both types are sown together and the 'blind' seeds may give rise to apothecia from which ascospores are discharged, infecting the ovaries of rye-grass and darnel and thus leading to the formation of more 'blind' seeds. Even some rust fungi may be 'seed-borne' by being included in a commercial seed sample. When seed from a rusted flax crop is harvested teliospores of *Melampsora lini* attached to small fragments of fruit-walls and leaves get mixed with the seed (Muskett and Colhoun 1947). Next season these fragments are scattered with the seed and in due course the teliospores germinate to produce basidiospores that are discharged and infect the young flax plants.

The various types of seed-borne fungi have been illustrated by reference to a few scattered examples, but there are very many more, and plant pathologists are becoming increasingly aware of the importance of this type of dispersal in agricultural practice. How far seed dispersal is of importance for fungal pathogens attacking plants under natural conditions is not yet known. Harley (1950) reported the examination of seeds of a number of wild and garden plants removed with due precautions from unopened capsules and left overnight in sterile water. He says: 'Excluding Cistaceae and Ericaceae, in all of which fungi occurred, eleven out of twenty-five species were seen to bear hyphae on some of the seeds.'

A consideration of seed-borne fungi shows all types of association, from the most casual one in which the spores simply adhere to the surface of the grain (e.g. *Tilletia caries*) to the most highly organized where the fungus has ceased to sporulate and relies entirely for its dispersal on a seed-borne mycelium (e.g. the endophyte of *Lolium*).

From the point of view of disease prevention the location of the parasite in the seed is of great importance. Surface infection, with the spores or resting mycelium on or near the surface of the seed or seed-like fruit, is the commonest type and can usually be tackled successfully by seed treatment with a suitable fungicide. However, where deeper tissues are affected disinfection cannot be achieved in this way. Luckily, deep-seated infection is rarer. In *Ustilago nuda*, as already mentioned, seed sterilization can be effected by hot-water treatment, but this is not a method of general application. In anthracnose of beans, for example, the only efficient method of control is to avoid the use of seed from a diseased crop.

The types of fungi that are normally seed-borne, in the sense that the fungus is actually in or on the seed, are rather limited. Amongst Basidiomycetes they are smuts. Otherwise they are mostly Fungi Imperfecti or Ascomycetes in the 'imperfect' stage. The main groups of truly obligate parasites (i.e. Peronosporaceae, Erysiphales, and Uredinales) are not usually seed-borne. It is interesting to note that in all these there is not only a well-developed airborne stage, but also an efficient hibernating stage. Seed-borne infection as often as not serves rather as a method of hibernation than of dispersal.

The plant pathologist often extends the conception of seed-borne fungi to include those carried by vegetative reproductive units or 'planting units' such as tubers, bulbs, and runners (Muskett 1950). But although dispersal of this nature may be of great practical significance, it has little importance in nature.

16. DISPERSAL IN AQUATIC FUNGI

THE great majority of fungi are terrestrial, probably less than 2 per cent being submerged aquatics. These are, however, deserving of special consideration since problems of dispersal through an aqueous medium are somewhat different from those of aerial dissemination.

From the phylogenetic point of view there would seem to be two essentialy different types of aquatic fungi among both freshwater and marine species. First are those whose whole evolutionary story appears to have been an aquatic one. They bear the stamp of primary aquatics in possessing zoospores. To this class belong the Chytridiomycetes, with a zoospore having a posterior flagellum of the whip-lash type, and the Oomycetes with biflagellate zoospores each with one flagellum of the whip-lash type directed backwards and the other of the 'tinsel' type in the forward position. There is also a very small group (Hyphochytriomycetes) in which the zoospore has a single anterior 'tinsel' flagellum (Fig. 16.1).

Secondly there are the aquatic fungi which seem most likely to have arisen from terrestrial ancestors. They include many Ascomycetes, a

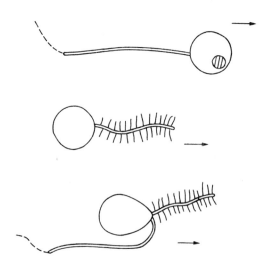

FIG. 16.1. *Top*: chytrid zoospore. *Middle*: zoospore of *Rhizidiomyces* *Bottom*: zoo-spore of *Saprolegnia*. Arrows indicate direction of swimming. × *c*. 1200. Somewhat diagrammatic but based on Couch (1941).

considerable array of 'imperfect' forms, both Hyphomycetes and Sphaerop-
sidales, but so far only two completely submerged members of the
Higher Basidiomycetes have been described.

As with aerial fungi three episodes: spore release, transport (dispersal)
through the water, and deposition can usually be recognized, although it
must be conceded that little is known about actual dispersal. This is
largely because, unlike the position with terrestrial fungi, there is little
economic incentive to study this problem.

Members of Saprolegniales are probably the most familiar water moulds,
so often forming a dense growth on dead fish in lakes and canals, some
even being implicated in death of the fish. The process of spore maturation and release has been studied by Gay and Greenwood (1966) in
Saprolegnia, using cinematography supported by electron micrographs
of sections of the critical stages. The zoosporangium (Fig. 16.2 and
Plate VIII) is usually cigar-shaped and cut off by a cross-wall from a
hypha projecting into the water. At a fairly early stage there is a single
central vacuole. The membrane delimiting this from the thickish layer
of peripheral protoplasm is the tonoplast, while the plasmalemma is at
the interface against the cell-wall. In the formation of zoospores cleavage
furrows strike outwards from the vacuole and when these reach the
plasmalemma the process is complete. At this stage the limiting plasma
membrane of each zoospore is made up of a portion of plasmalemma and
a portion of tonoplast. All the protoplasm is used up in zoospore cleavage.
At the moment when the ridges of tonoplast join the plasmalemma,
there is no longer a plasma membrane separating the vacuolar sap from
the wall. Thus there is no longer a system to maintain the turgor of the
whole sporangium and the solutes of the former vacuole are presumably
free to diffuse into the surrounding water. This seems to explain, first,
the sudden retraction of the zoosporangium, which loses about 10 per
cent of its volume, and secondly the bulging of the previously flat cross-
wall at the base of the sporangium. Soon afterwards the zoospores may be
seen to be oscillating vigorously, jostling one another. A little later the tip
of the apical papilla of the zoosporangium breaks down and the pear-
shaped zoospores escape. This they do in a rush, most escaping in a
second or two as if they are being squeezed out as from a bursting ascus.
However, the zoosporangium at this stage would seem to have lost its
turgidity, so it is difficult to explain its rapid evacuation. Although most
spores escape in the first few seconds following dehiscence, a few lag
behind, and these move in a more leisurely fashion. Each passes up the
middle of the sporangium, apparently swimming. The blunt end is
forward and the more pointed end, bearing the two flagella, is behind.
Once outside the sporangium, however, the zoospore reverses and swims
with the pointed end forward, the tinsel flagellum apparently lashing
in front and the whip-lash one tending to trail behind.

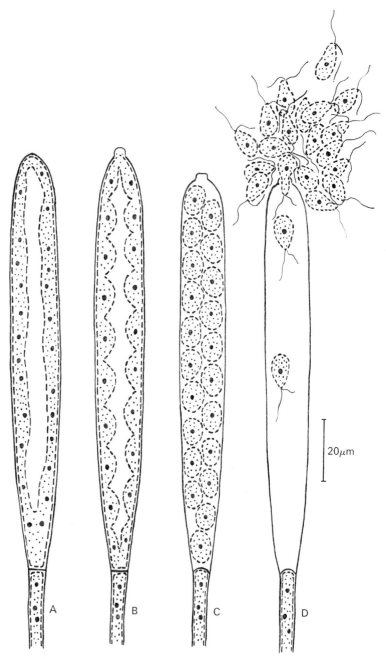

20μm

FIG. 16.2. Diagram representing maturation of sporangium and liberation of zoospores. Plasmalemma and tonoplast shown by dashed lines. Note loss of 10 per cent of volume between (B) and (C) and change of form of basal wall in (C). At 20°C about half-an-hour elapses between (A) and (B) and 5 min between (B) and (C). Only a few seconds separate (C) and (D). The scale is very approximate. Based on Gay and Greenwood (1966) and on film by Dr. Greenwood referred to in their paper.

Although in *Saprolegnia* the zoospores escape in such a rush that it is difficult to believe that only their own motility is involved, there are many aquatic fungi in which liberation of the zoospores from the sporangium is clearly the result of their individual motility. This is certainly the position in many chytrids.

A special condition is presented by species of *Pythium* in which, characteristically, the protoplasm of the sporangium escapes in one mass contained in a very thin membrane. In this vesicle final zoospore delimitation occurs and when the membrane is finally ruptured the spores are free to escape. The details of this interesting process have been investigated by Webster and Dennis (1967). As with studies of *Saprolegnia*, a cinematograph record has greatly helped in a study of the changes that occur.

In *Pythium middletonii* the nearly mature spherical sporangium (*c.* 30 μm diam) is filled with dense protoplasm and has a tubular papilla (20 μm long $\times$ 7 μm diam) with a terminal cap having a rather thicker wall than the rest of the papilla (Fig. 16.3). At maturity this cap is inflated into a very thin-walled vesicle into which all the protoplasm passes, and there it cleaves into separate zoospores which, by their vigorous movement, rupture the vesicle and escape.

The initial stages of the passage of the protoplasm into the vesicle seem to be related to a shrinkage of the sporangium to about half its original volume. Thereafter, the protoplasm begins to separate from the inner surface of the sporangium so that its wall pressure must at this stage cease to be a factor. Neither does any pressure of the vesicle wall seem to be involved, for very soon the external protoplasm does not completely fill it. Webster and Dennis point out that at this stage there is a situation in which two droplets of protoplasm are in communication through a tubular connection. If the outer droplet were larger than the inner, surface tension might be expected to lead to all the protoplasm flowing into the larger one. It is pointed out that the situation is comparable with the physical model that can be used to demonstrate that the excess pressure inside a soap bubble is inversely proportional to its radius, so that when the interior of a small bubble is put, by means of a glass tube, in communication with that of a larger one, the smaller collapses with a corresponding growth of the larger (Fig. 16.4). A situation involving drops rather than bubbles is fundamentally similar. However, in the *Pythium* sporangium, at the time when wall pressure ceases to drive protoplasm into the vesicle, the outer droplet of protoplasm has a diameter somewhat *less* than that of the droplet remaining in the sporangium. If, however, the surface tension of the outer protoplasmic droplet were less than that of the inner, the outer one could still grow and acquire all the protoplasm. Webster and Dennis (1967) make reasonable suggestions about how this might occur.

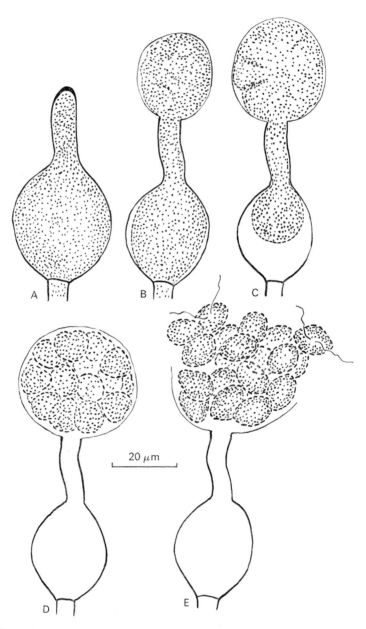

Fig. 16.3. *Pythium middletonii*. Stages in sporangium discharge. (A) 6 min before discharge (note thickened cap of papilla); (B) a few seconds after the start of discharge (protoplasm just separating from sporangial wall); (C) 40 s after discharge began; (D) about 15 min later; (E) after a further 2 min. From stills from a film. From photographs by Webster and Dennis 1967.

An alternative explanation, as Webster and Dennis indicate, is that amoeboid movement of the protoplasmic mass might explain migration into the vesicle. This would involve a different type of mechanism since students of amoeboid movement now reject the idea that surface tension plays an essential role in this process.

In these primitive aquatic fungi once the zoospore has escaped from the sporangium, its transport is in part due to its own motility and in part to the motion of the water itself. It is hardly appropriate to discuss in any

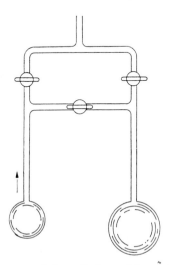

FIG. 16.4. Apparatus, consisting of glass-tubing and two-way taps, to illustrate that excess pressure in a bubble is inversely proportional to its size. Bubbles of unequal size are blown at the end of each tube with the appropriate tap open and the middle one closed. Then, with the lateral taps closed, the two bubbles are put into communication by opening the middle tap. The smaller bubble shrinks to nothing and the larger increases proportionately. After Webster and Dennis 1967.

detail the movements of the flagella that are responsible for motility for this would lead into an area of study involving flagellates generally.

So far as the aquatic Phycomycetes are concerned the most distinguished contribution to the study of zoospore swimming is that of Couch (1941) who examined motility in a considerable range of genera belonging to all the zoosporic orders. He found that in the single forward-projecting tinsel flagellum of *Rhizidiomyces* (Hyphochytriomycetes) the wave motion starts near the end and travels backwards to the spore and this apparently has the effect of dragging the spore forward. On the other hand, in the single posterior flagellum of a chytrid zoospore, the wave starts at the spore and moves towards its tip driving the spore forward after the manner of an animal sperm. According to Couch, in the biflagellate zoospore (e.g. in *Saprolegnia*) the two types of flagellum behave as they

(18a)

do in the uniflagellate zoospores, the tinsel flagellum, projecting forwards, and the trailing whip-lash flagellum both contributing to the movement of the spore in a pull-and-push operation. However, McKeen (1962) came to the conclusion that in biflagellate Phycomycetes the anterior flagellum serves only as a rudder and does not move, propulsion being due to the activity of the posterior whip-lash flagellum.

From the point of view of dispersal certain features of zoospores are important. First, their length of life, secondly, the rate at which they swim, and thirdly their ability to come to rest on a suitable substratum. On the first two there is very little firm information. Many zoospores certainly keep swimming for several hours and some, perhaps, for several days. No doubt the duration is affected by conditions, particularly temperature and the availability of a surface on which to settle. So far as the rate of progress is concerned, it is possible to calculate the speed of the zoospores of *Pythium aphanidermatum* from the data of Royle and Hickman (1964*a*, *b*). It is about 1·0 cm/min. The present author has observed a value of about 0·25 cm/min for zoospores of *Chytridium olla*. If these rates are at all typical, it is clear that, in general, they are likely to be small in relation to the general movement of the water in which the zoospores are suspended. Further, few zoospores maintain a steady direction for a significant time. Most show sudden and haphazard changes of direction from time to time, and some (e.g. those of *Chytridium olla*) even have the habit of swimming in small circles.

We have seen that the biological advantage of motility may lie in providing a mechanism of escape from the zoosporangium, but probably the major value lies in the power the zoospore possesses of 'selecting' its substratum to some extent by virtue of its taxisms.

A few water moulds have zoospores that are positively phototactic. An example is *Polyphagus euglenae*, which is parasitic on *Euglena*, itself phototactic. The result is that the fungal zoospores tend to accumulate in the same illuminated regions as do the host cells, and no doubt this increases the chances of parasitic attack.

Probably the zoospores of most water moulds are chemotactic. This response has been known for a long time in Saprolegniaceae, and last century was the subject of extensive researches by Pfeffer (1884). More recently the problem has been investigated by Fischer and Werner (1958). Zoospores of *Saprolegnia* spp. are strongly attracted by such natural 'baits' as the bundle of muscle fibres exposed in the detached leg of a fly. They are not attracted by protein freed from soluble substances by dialysis, but small quantities of amino acids are active as also are certain salts particularly the chlorides of potassium and sodium. Mixtures of amino acids and these salts are especially efficient in evoking a chemotactic response. These are just the kinds of solutes that might be expected to diffuse from a dead fly or fish.

In an investigation of chemotaxis in the water mould *Allomyces* Machlis (1969) found that zoospores were attracted by casein hydrolysate, the activity being traceable to the combined action of the amino acids leucine and lysine.

Chemotaxis of zoospores has also been studied in relation to certain soil-borne fungi belonging to the two important genera *Pythium* and *Phytophthora*. In this chapter we are concerned essentially with truly aquatic fungi, but soil too has its aquatic phase and, indeed, the boundary between soil and aquatic fungi is not easy to draw. Dukes and Apple (1961) found that zoospores of *Phytophthora parasitica* were attracted to the roots of a large range of species and accumulated particularly around small fresh wounds in the roots. It was also discovered that the zoospores were attracted by certain sugars (sucrose, dextrose, fructrose, and maltose) but not by others (lactose and galactose). Casamino acids were also attractive. Work along these lines was carried further by Royle and Hickman (1964) using *Pythium aphanidermatum*. They too found that zoospores were attracted to roots and especially to wounds where sap was diffusing out. In the intact root attraction was to the subapical regions. They analysed the attraction of zoospores to extracted sap incorporated in agar in small capillary tubes immersed in a shallow layer of zoospore-containing water. In the water remote from the tubes zoospores swam in the form of an open corkscrew, and on coming near the open end of a capillary tube containing an attractive diffusate the general direction of swimming was towards the agar. On very close approach the rate of swimming was considerably slowed and the path of each zoospore because modified so that it performed looping movements with frequent and random changes of general direction (Fig. 16.5). The net effect of these two responses was an accumulation of zoospores around the mouth of the agar-filled capillary.

In summary, the view may be hazarded, on rather inadequate evidence, that motility of zoospores is probably of some importance in spore liberation in certain water moulds, of comparatively little value in actual dispersal, but possibly of considerable significance in landing on a suitable substratum.

Turning to those aquatic fungi that may well have been derived from terrestrial ancestors, we may consider first the 'imperfect' forms, particularly the conidial species classified as members of Moniliales or Hyphomycetes (Ingold e.g. 1942, 1966). In the freshwater habitat these occur abundantly, particularly as saprophytes on submerged decaying leaves of dicotyledonous trees and shrubs of all kinds in well-aerated waters. The babbling brook lined with willows or alders, or flowing through an oak wood, provides an ideal habitat for these fungi, especially in late autumn and winter when dead leaves may become very abundant in the stream bed. The branched septate mycelium ramifies in the dead leaf tissue and the conidiophores project into the water and there liberate their spores. The spore production, liberation, transport, and eventual deposition, occur below water.

The outstanding feature of these fungi is that their spores are mostly not of a conventional kind. Many have branched spores and of these the

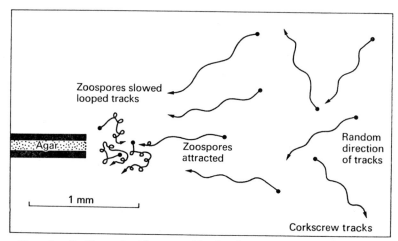

Fig. 16.5. *Pythium aphanidermatum.* Tracks of zoospores in shallow water in relation to an open capillary tube containing pea-root-extract agar. Each track shows movement of a zoospore over a period of 4 s, the dot indicating the position at the start and the arrow-head at the end of that period. Remote from the agar zoospores follow open corkscrew tracks unrelated to the exposed agar surface; nearer it the tracks are oriented towards the agar; closer still zoospores are slowed down and have looped tracks. Tracks shown here in two dimensions are really three-dimensional. Diagram based on photographs, figures, and description by Royle and Hickman (1964).

kind consisting of four arms diverging from a point is particularly common; less often the spore is very elongated and frequently sigmoid with a curvature lying in more than one plane. In a suitable stream, especially after heavy rain, cakes of semi-permanent foam collect below a waterfall or rapids. If such foam is collected, allowed to settle, and then a drop is examined under the microscope, it will be seen, particularly in autumn, to contain numerous spores of aquatic Hyphomycetes belonging to many species. The foam seems to act as a spore-trap retaining spores that happen to be brought to the surface by the turbulence of the water.

A special feature of the tetra-radiate spore is that there are so many ways in which it can develop.

Students of conidial fungi recognize a number of developmental patterns and, particularly, see a fundamental distinction between aleuriospores (sometime called terminal thallospores) and phialospores. In the former the spore primordium is delimited from the end of a hyphae by a cross-wall. Continued development leads to the maturation of the spore which finally separates from its conidiophore. Following this a new spore is not normally produced at once from the vacated conidiophore, or, if it is, the spore is not formed at exactly the same level as was the first. On the other hand the phialospore is produced from a special cell of the conidiophore, the phialide, usually narrower at its apex and often at its base than in its middle region. From the apex of the phialide the spore is produced, and it is separated by a cross-wall only when it is fully grown. Further, after its completion a new spore is usually immediately formed at the end of the phialide.

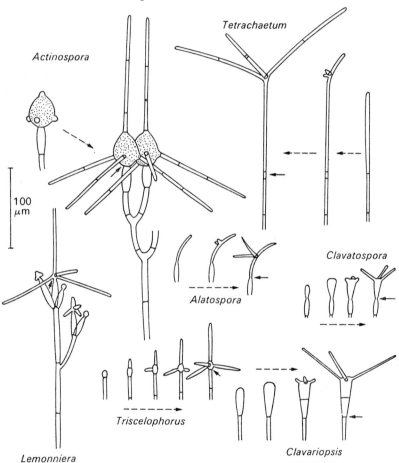

FIG. 16.6. Aquatic Hyphomycetes with tetraradiate spores indicating how development occurs. Relation of developmental stages indicated by arrows with 'dashed' shafts. Species figured: *Actinospora megalospora, Tetrachaetum elegans, Lemonniera aquatica, Alatospora acuminata, Clavatospora longibrachiata, Triscelophorus monosporus,* and *Clavariopsis aquatica.* Short arrows show point of attachment of mature spore.

Among the fungi with tetra-radiate spores are types where the spore is attached to its conidiophore by the tip of one of its arms, and types where the spore is attached near the point of divergence of the four arms (Fig. 16.6). Species belonging to some genera of the former kind may be considered first. In *Clavariopsis aquatica*, with aleuriospores, and *Clavatospora longibrachiata*, with phialospores, the spore primordium is a narrowly obconcial structure from the apex of which three arms arise and grow out simultaneously. In both fungi the liberated tetra-radiate spore consists of one fat arm and three much thinner ones. In *Tetrachaetum elegans*, with aleuriospores, and *Alatospora acuminata*, with phialospores, the two arms of the mature tetra-radiate spore are derived from an original axis bent in the middle and two laterals that arise simultaneously where the axis is bent.

Considering species where the spore is attached near the region of divergence of the four arms, *Lemonniera aquatica* has phialospores in which the four arms develop simultaneously. In *Actinospora megalospora* development is rather similar, but the spores are aleuriospores, as they are in *Triscelophorus monosporus* in which the arms arise in succession, not simultaneously.

A number of other patterns of development of the tetra-radiate spore could be cited by reference to the genera *Articulospore*, *Tetracladium*, *Gyoerffyella*, and *Campylospora*, and it is clear that this kind of spore has probably been developed independently along a number of quite separate lines. Convergent evolution seems to have occurred. This strongly suggests that a spore of this kind has a special survival value in the environment in which these fungi occur.

There are two plausible advantages of the tetra-radiate spore. Spores of aquatic fungi generally have a density slightly greater than 1·0 and therefore sink in still water. However, a tetra-radiate spore probably sinks more slowly than a spherical one of the same volume, and might, therefore, remain in suspension longer with greater chance of effective dispersal. This idea is in keeping with the generally held view that the biological advantage of the long processes and spines so common on plankton algae is related to maintaining these organisms in the photic zone. However, the rate of settling of an aquatic spore of any form is likely to be very small in relation to water movements, particularly in streams. Again, if fungal spores had tended to evolve in the direction of developing devices reducing their rate of sedimentation, such devices might be expected more in aerial species than in aquatic ones, since the rate of fall of any spore in air is so much greater than in water.

The second possibility is that the tetra-radiate form is associated with ultimate anchorage, and, indeed, the coming to rest of a spore on an appropriate substratum, such as a dead leaf or a twig, is a very real problem in a stream. It might reasonably be expected that a tetra-radiate spore, in comparison with a spherical or ovoid one, would be more readily impacted on an underwater object, but clearly experimental verification is desirable. This has been provided by Webster (1959), who used a horizontal water-tunnel through which water containing a known concentration of spores could be passed at controlled speeds (Fig. 16.7). Into this tunnel a vertical glass rod was inserted as a spore-trap. Later it was removed, its leading surface scanned under the microscope, and counts of deposited spores were made. The efficiency of

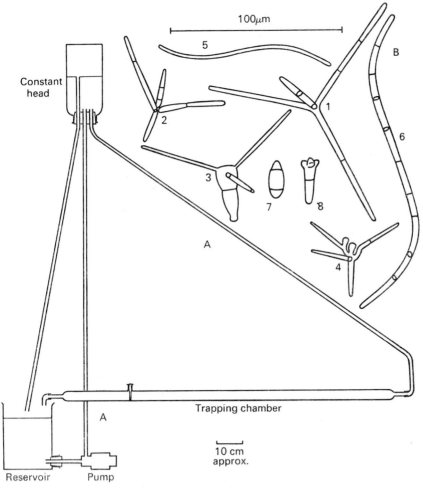

FIG. 16.7. Impaction of spores of aquatic Hyphomycetes. (A) 'water-tunnel' apparatus used by Webster (1959). (B) spores of species indicated in Table 16.1. (1) *Lemonniera aquatica*; (2) *Articulospora tetracladia*; (3) *Clavariopsis aquatica*; (4) *Tetracladium marchalianum*; (5) *Flagellospora curvula*; (6) *Auguillospora longissima*; (7) *Dactylella aquatica*; (8) *Heliscus lugdunensis*.

impaction of the spores could be calculated, and it was found, for example, that this was very much higher for the tetra-radiate spores of *Articulospora tetracladia* than for spores of a more conventional form such as occur in *Heliscus lugdunensis* and *Dactylella aquatica* (Table 16.1).

An interesting feature of the spores of aquatic Hyphomycetes is that while suspended in water they do not germinate. However, on coming to rest on a solid object, germination rapidly occurs, usually involving the production of a hypha from the tip of each arm.

TABLE 16.1

Trapping efficiency (per cent) of some aquatic hyphomycete spores at three water speeds. Each figure is mean of three experiments. Extracts from much larger table by Webster 1959.

	15 cm/s	25 cm/s	35 cm/s	Mean
TETRA-RADIATE SPORES				
Articulospora tetracladia	0·35	0·51	0·33	0·40
Tetracladium marchalianum	0·28	0·37	0·27	0·31
Clavariopsis aquatica	0·3	0·4	0·3	0·33
Lemonniera aquatica	0·35	0·13	0·04	0·17
LONG, SIGMOID SPORES				
Anguillospora longissima	0·11	0·1	0·1	0·10
Flagellospora curvula	0·06	0·14	0·14	0·11
OTHER TYPES OF SPORE				
Heliscus lugdunensis	0·004	0·009	0·007	0·007
Dactylella aquatica	0·004	0·009	0·008	0·007

Emphasis has been laid on those species with tetra-radiate spores. However, a number of aquatic Hyphomycetes have branched spores that are not tetra-radiate (e.g. *Dendrospora erecta* and *Polycladium equiseti*), quite a number have very elongated spores with curvature in more than one plane (e.g. *Flagellospora curvula*, *Anguillospora longissima*, and *Lunulospora curvula*), and a few have spores that are round or oval (e.g. *Margaritispora aquatica* and *Dactylella aquatica*). So far as the elongated spores are concerned, the same type of story can be told as for the tetra-radiate spores and probably this form of spore can again be related to impaction. In Webster's table, spores of *Anguillospora* and *Flagellospora* occupy an intermediate position between the tetra-radiate spores and those of *Heliscus* and *Dactylella*.

It must be remembered that elongated sigmoid and tetra-radiate spores are not limited to the aquatic environment, but they tend to be so strongly represented there as to constitute a major characteristic of the water spora. One well-known fungus *Tetraploa aristarta* with tetra-radiate spores has repeatedly been reported in the air-spora. This fungus, however, occurs particularly on the aerial parts of reed-swamp plants, but it is also capable of growing and sporulating below water and its spores have frequently been seen, associated with the spores of true aquatic Hyphomycetes in foam samples.

There are also numerous sub-aerial fungi with thread-like spores, but again spores of this kind tend to be particularly richly represented in the water spora.

The tetra-radiate aquatic propagule is by no means limited to fresh-water aquatic Hyphomycetes of submerged decaying leaves and twigs. It is also found in a few pycnidial fungi, for example, in *Robillarda phragmites*, which occurs on submerged stalks of *Phragmites*. Spores of *Robillarda* spp. are frequently to be seen in the foam of rivers and streams in many parts of the world. However, perhaps the most striking occurrence of the tetra-radiate spore outside 'imperfect' fungi is in *Digitatispora marina*. This seems to be the only hymenomycete known to produce a submerged sporophore. It occurs on wood in the sea and was originally described by Doguet (1962*a*, *b*) from the coast of Northern France and later was also discovered in the Pacific (Kohlmeyer 1963). Structurally it appears to be close to *Corticium*, with a resupinate sporophore having a smooth hymenium. Each basidium is rather elongated and bears a crown of four basidiospores. Its cytology agrees exactly with the normal picture of basidium development (Doguet 1962*b*). However, there are striking differences from the aerial Hymenomycetes. First, the basidiospores are sessile, being without sterigmata. Secondly, each spore is tetra-radiate bearing a remarkable resemblance to the kind of spore so often found in aquatic Hyphomycetes (Fig. 16.8). Perhaps even more remarkable has been the discovery of Doguet (1967) that *Nia vibrissa* is a marine gasteromycete. It grows on wood and the spherical sporophore (2–3 mm diam) has a thin peridium around the gleba. In this are the basidia, each with a filamentous base and a spherical apex bearing 5–8 basidiospores. The body of the basidiospore is ovoid and is attached to the basidium by a short stalk, but in addition there are five or more long filamentous appendages (Fig. 16.9). At maturity the basidia break down and the sporophore then contains thousands of basidiospores immersed in mucilage. No doubt it is by the swelling of this that the peridium is ruptured setting the spores free to drift away.

It is also of interest to note that the tetra-radiate aquatic propagule has also been evolved outside the fungi in a group of brown seaweeds, namely Sphacelariales. A number of *Sphacelaria* spp. reproduce by propagules that are deciduous lateral branches of limited growth and definite form, and in some species these are tetra-radiate.

In striking contrast to Basidiomycetes, the higher Ascomycetes are

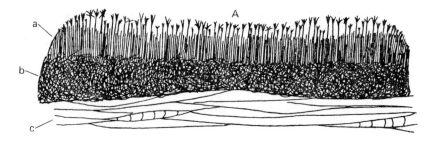

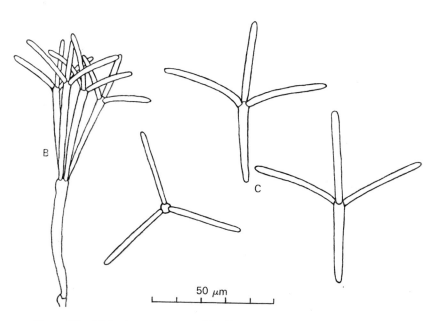

FIG. 16.8. *Digitatispora marina*. (A) diagrammatic section through the sporophore: (*a*) hymenium, (*b*) general tissue of sporophore, (*c*) decaying wood. After Kohlmeyer (1963). (B) basidium. (C) three liberated spores. After Doguet (1962*a*).

moderately common both in fresh water and in the sea. In the fresh-water habitat both Discomycetes and Pyrenomycetes are regularly to be found on submerged dead reed-swamp stalks (e.g. of *Phragmites, Schoeno-plectus,* and *Equisetum*) and to a lesser extent on sunken twigs and branches. In the sea it seems that only Pyrenomycetes occur and these are mostly to be found on wood that has been submerged for a long time. A few species also parasitize the larger seaweeds. Perhaps the best known of these is *Mycosphaerella ascophylli*, which occurs in *Ascophyllum nodosum*.

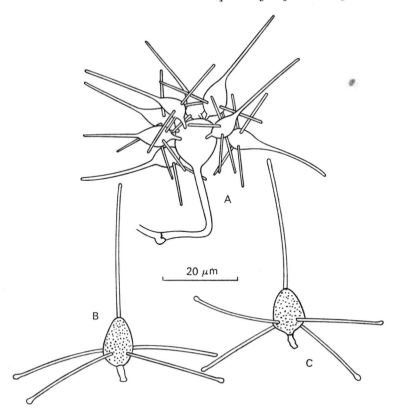

FIG. 16.9. *Nia vibrissa.* (A) basidium with 8 basidiospores; note clamp-connection at its base. (B and C) two liberated basidiospores. After Doguet (1967).

Indeed in that abundant seaweed a specimen has never been found to be without this systemic endophyte which does no damage to its host. Sporulation is limited to the receptacles in which the conceptacles are intermixed with the much smaller and black perithecia. Ascospore liberation from the fungus and the release of eggs from *Ascophyllum* occur in the same period, but the mechanism that brings about the invariable association of the two organisms is not clear (Webber 1967).

Although there are many aquatic Ascomycetes, their number probably does not exceed a few hundred, in contrast to the tens of thousands of terrestrial species. It seems most likely that the present flora of aquatic species has been derived essentially from terrestrial ancestors, although in the still dimmer and more distant past the higher Ascomycetes generally may have had aquatic forebears. However, to return from these realms of fantasy, it is to be remarked that an ascus fruit-body is essentially a mechanism capable of functioning in water, indeed terrestrial forms can have their exposed hymenia wetted by rain without harm. On the other hand, the normal hymenomycete hymenium of basidia is ruined if wetted.

The freshwater aquatic discomycete functions just as does a cup-fungus on land, except that, because of the much greater viscosity of water as compared with that of air, the distance of discharge is greatly reduced. However, below water this distance is of no real significance for dispersal. All that is necessary is that the spores shall be freed from the parent body. Generally the submerged discomycetes show little indication of special adaptations to their habitat. Many species have spores very little different from those of land species, but some have long, thread-like spores (e.g. *Vibrissea truncorum* and *V. guernisaci*). As we have seen, the thread-like spore is a feature of aquatic life. Not only does it occur in a considerable number of Ascomycetes, both freshwater and marine, but we have seen it in several types of aquatic Hyphomycetes. Further, it is very suggestive that in the only flowering plant in which pollination occurs below water, namely *Zostera*, the pollen grains instead of being spherical are very long and thread-like.

On the whole, it is more often possible to see in freshwater Pyreno-mycetes suggestions of adaptation to aquatic life. Two outstanding examples, both saprophytes of reed-swamp plants, may be cited. In the first, *Loramyces* spp., the liberated 2-celled ascospore has a long 'tail' and a 'head' surrounded by mucilage (Fig. 2.1A). As the spore glides, head foremost, with a current of water, it readily impacts on a submerged reed and becomes stuck. In the second, *Pleospora scirpicola*, the large ascospore is contained in an elongated 'slug' of mucilage. No doubt this also has a significance in relation to underwater impaction. However, this theme of ascospore modification, apparently related to impaction, reaches its fullest expression in the marine Pyrenomycetes.

During the past twenty years considerable attention has been paid to the higher Ascomycetes growing on wood submerged in the sea. They all appear to be Pyrenomycetes and most of the genera are peculiar to the sea and are not merely marine species of terrestrial genera. However, the outstanding feature of these fungi is that their spores mostly have various types of sticky appendages which can be interpreted as of significance in impaction (Fig. 16.10).

There is some scattered information about the actual process of spore discharge from aquatic perithecia. As with apothecia, it is essentially like that in terrestrial species, although the actual distance of discharge is insignificant. The process has been described in the freshwater *Ophiobolus typhae* in which the detached-ascus type is found resembling *Ceratostomella* (see p. 46). Again Kohlmeyer (1960) has illustrated spore release in *Herpotrichiella ciliomaris* in which more than one ascus can protrude at a time from the rather large ostiole (Fig. 16.11). In some other marine Pyrenomycetes, as in the terrestrial *Chaetomium*, the asci break down

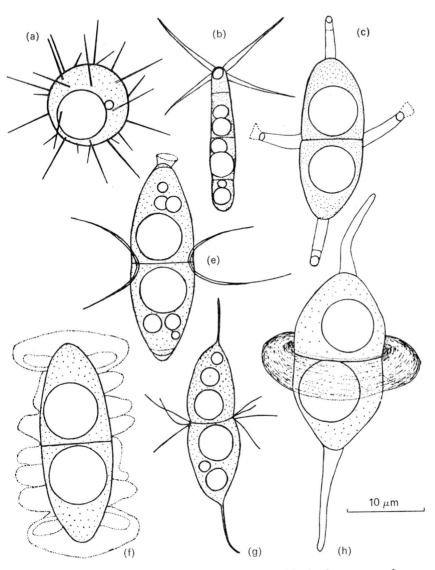

FIG. 16.10. Ascospores of marine Ascomycetes: (*a*) *Amylocarpus encephaloides*, (*b*) *Torpedospora radiata*; (*c*) *Ceriosporopsis calyptrata*; (*d*) *Halosphaeria mediosetigera*; (*e*) *Remispora maritima*; (*f*) *Peritrichospora integra*; (*g*) *Halosphaeria torquata*. After Kohlmeyer (1962?).

within the perithecium and the ascospores ooze out in mucilage (Fig. 16.12).

A great lack in the dispersal picture of aquatic fungi is that almost nothing is known quantitatively about the water spora, in striking contrast to our considerable knowledge of the air spora. Willoughy (1962), however, using an indirect method of estimation, found concentrations

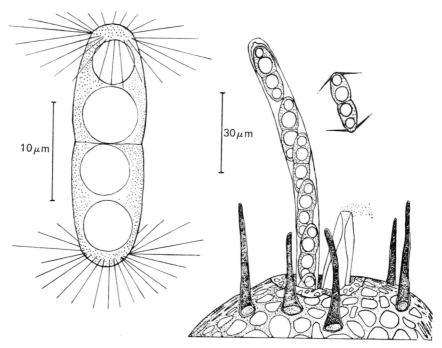

Fig. 16.11. *Herpotrichiella ciliomaris. Right:* top of perithecium with two asci protruding through the ostiole; the contracted ascus on the right has just emptied; that on the left is at a stage just before spore liberation. *Left:* a liberated spore more highly magnified. After Kohlmeyer (1960).

of up to 5000 saprolegniaceous propagules (probably spores) per litre at the lake margin of Windermere, although the values were usually much lower. There was a marked seasonal periodicity with values high during autumn but seldom exceeding 100 propagules per litre during spring and summer. Concentrations were much lower in the middle as compared with the margin of the lake.

Because the spores of aquatic Hyphomycetes are so easy to identify, it should be possible to examine their concentrations in streams throughout the year, but studies of this kind have yet to be made. So far as the sea is concerned it might be very difficult to study the marine spora mainly

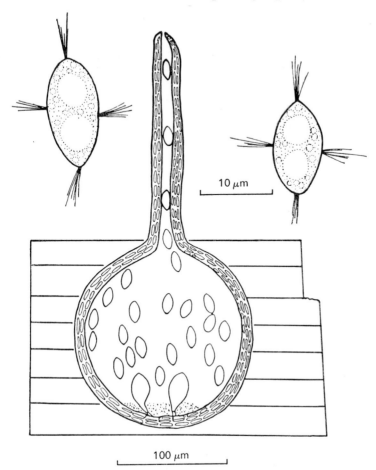

10 μm

100 μm

FIG. 16.12. *Nautosphaeria cristominuta*. Vertical section of perithecium with base immersed in wood and neck projecting; intact asci at base and freed spores some of which are passing out through the neck canal. Also two liberated ascospores at higher magnification. After Gareth Jones (1964).

because the concentration of spores is likely to be so low. In the sea and in lakes spore production is unlikely to be on the grand scale that is such a feature of the terrestrial habitat.

In freshwater aquatic Hyphomycetes, it has been suggested that the dispersal process involving production, liberation, transport, and deposition of spores occurs below water, and the same may be said of submerged species of other groups. This picture is, however, true only for one isolated water system. Nevertheless, it has become apparent that in aquatic Hyphomycetes most species are remarkably widespread. Thus *Lemonniera aquatica, Clavariopsis aquatica, Tetracladium marchalianum,*

Triscelophorus monosporus, *Tricladium gracile*, *Campylospora chaetocladia*, and *Lunulospora curvula* all occur in Europe, Asia, Africa, Australia, and America. This raises the question of transport from one separate freshwater situation to another, and no answer based on real evidence can be given. It seems most unlikely that the spores as such are essentially involved. They are thin-walled delicate structures that probably have little ability to withstand dessication. However, in many species the mycelium, or part of it, becomes dark and thick-walled and sometimes intercalary cells constitute definite chlamydospores. Very probably long-distance dispersal of these fungi, as in so many freshwater plants, is through the agency of waterfowl. Pieces of decaying leaf containing the resistant mycelium may well be the effective units in this kind of dispersal.

BIBLIOGRAPHY

ADAMS, K. F., HYDE, H. A., and WILLIAMS, D. A. (1968) Woodlands as a source of allergens with special reference to basidiospores. *Acta allerg.* **23**, 265–81.

AHMADJIAN, V. (1967) *The lichen symbiosis.* Waltham, Mass.

AINSWORTH, G. C. (1961) *Ainsworth and Bisby's dictionary of fungi.* London.

—— (1962) Longevity of *Schizophyllum commune.* II. *Nature, Lond.* **195**, 1120–1.

ALASOADURA, S. O. (1963) Fruiting in *Sphaerobolus* with special reference to light. *Ann. Bot.* **27**, 125–45.

ANATASOFF, D. (1920) Ergot of grains and grasses. U.S. Department of Agriculture, Bureau of Plant Industry. Mimeographed publication, pp. 1–127.

—— (1925) The *Dilophospora* disease of cereals. *Phytopathology* **15**, 11–40.

ANDERSEN, A. A. (1958) New sampler for the collection sizing, and enumeration of viable airborne particles. *J. Bact.* **76**, 471–84.

ARX, J. A. VON (1968) *Pilzkunde.* Verlag von J. Cramer.

ASHBY, S. F. and NOWELL, W. (1926) The fungi of stigmatomycosis. *Ann. Bot.* **47**, 69–83.

AUSTIN, B. (1968*a*) Effects of airspeed and humidity changes on spore discharge in *Sordaria fimicola. Ann. Bot.* **32**, 251–60.

—— (1968*b*) An endogenous rhythm of spore discharge in *Sordaria fimicola. Ann. Bot.* **32**, 261–78.

BAKER, G. E. (1966) Inadvertent distribution of fungi. *Can. J. Microbiol.* **12**, 109–12.

BAKER, H. G. (1947) Infection of species of *Melandrium* by *Ustilago violacea* (Pers.) Fuckel and the transmission of the resultant disease. *Ann. Bot.* **11**, 332–61.

BAKSHI, K. B. (1950) Fungi associated with ambrosia beetles in Great Britain. *Trans. Br. mycol. Soc.* **33**, 111–20.

BARKSDALE, T. H. and ASAI, G. N. (1961) Diurnal spore release of *Piricularia oryzae* from rice leaves. *Phytopathology* **51**, 313–17.

BENJAMIN, R. K. (1959) The merosporangiferous Mucorales. *Aliso* **4**, 321–433.

—— (1965) Addenda to 'The merosporangiferous Mucorales'. III. *Dimargaris. Aliso* **6**, 1–10.

BOCK, K. R. (1962) Dispersal of uredospores of *Hemileia vastatrix* under field conditions. *Trans. Br. mycol. Soc.* **45**, 63–74.

BOND, T. E. T. (1952) A further note on size and form in agarics. *Trans. Br. mycol. Soc.* **35**, 190–4.

BONDE, R. and SCHULTZ, E. S. (1943) Potato refuse piles as a factor in the dissemination of late blight. *Maine Agricultural Experimental Station Bulletin* N. 416; *Am. Potato J.* **20**, 112–18.

BOURDILLON, R. B., LIDWELL, O. M., and THOMAS, J. C. (1941) A slit asmpler for collecting and counting air-borne bacteria. *J. Hyg., Camb.* **41**, 197–224.

BOURSNELL, J. G. (1950) The symbiotic seed-borne fungus in Cistaceae. I. Distribution and function of the fugus in the seedling and in the tissue of the mature plant. *Ann. Bot.* **14**, 217–43.

BRANDT, W. H. (1953) Zonation in a prolineless strain of *Neurospora*. *Mycologia* **45**, 194–208.

BREFELD, O. (1877) *Botanische Untersuchungen uber Schimmelpilze*. III. Leipsig.

BRIDGER, D. H. (1967) Spore liberation in *Sporobolomyces* in relation to light. *Proc. Linn. Soc. Lond.* **178**, 49–58.

BRODIE, H. J. (1931) The oidia of *Coprinus lagopus* and their relations with insects. *Ann. Bot.* **45**, 315–44.

—— (1951) The splash-cup dispersal mechanism in plants. *Can. J. Bot.* **29**, 224–34.

—— (1956) The structure and function of the funiculus of the Nidulariaceae. *Svensk bot. Tidskr.* **50**, 142–62.

BRUCH, C. W. (1967) Microbes in the upper atmosphere and beyond. In *Airborne microbes, Proceedings of the seventeenth symposium of the Society for General Microbiology* (Editors P. H. Gregory and J. L. Monteith), pp. 345–74. Cambridge.

BUCHANAN, T. S., and KIMMEY, J. W. (1938) Initial tests of the distance of spread to and intensity of infection on *Pinus monticola* by *Cronartium ribicola* from *Ribes lacustre* and *R. viscosissimus*. *J. agric. Res.* **56**, 9–30.

BUCHWALD, N. F. (1938) Om Sporenproduktionens størrelse hos Tøndersvampen. *Friesia* **2**, 42–69.

—— and HELLMERS, E. (1946) Fortsatte Iagttagelser over Sporefaeldning hos Tøndersvamp (*Polyporus fomentarius* (L.) Fr.). *Friesia* **3**, 212–16.

BULLER, A. H. R. (1909, 1922, 1924, 1931, 1933, 1934, 1950) *Researches on fungi*, Vols. I–VII. London.

—— (n.d.) Unpublished chapters of *Researches* deposited at Royal Botanic Gardens, Kew.

BULMER, G. S., and BENEKE, E. S. (1964) Germination of basidiospores of *Lycoperdon* species and *Scleroderma lycoperdoides*. *Mycologia* **56**, 70–6.

BURDSALL, H. H. (1965) Operculate asci and puffing of ascospores in *Geopora* (Tuberales). *Mycologia* **57**, 485–8.

—— (1968) A revision of the genus *Hydnocystis* (Tuberales) and the hypogeous species of *Geopora* (Pezizales). *Mycologia* **60**, 496–525.

CALDIS, P. A. (1927) Etiology and transmission of endosepsis (internal rot) of the fruit of fig. *Hilgardia* **2**, 287–328.

CALLAGHAN, A. A. (1962) Observations on perithecium production and spore discharge in *Pleurage setosa*. *Trans. Br. mycol. Soc.* **45**, 249–54.

CARPENTER, J. B. (1949) Production and discharge of basidiospores of *Pellicularia filamentosa* (Pat.) Rogers on *Hevea* rubber. *Phytopathology* **39**, 980–5.

CARTER, M. V. (1957) *Eutypa armeniacae* (Hansf. and Carter), an airborne vascular pathogen of *Prunus armeniacae* in southern Australia. *Aust. J. Bot.* **5**, 21–35.

—— (1963) *Mycosphaerella pinoides*. II. The phenology of ascospore release. *Aust. J. agric. Res.* **16**, 800–17.

—— (1965) Ascospore deposition in *Eutypa armeniacae*. *Aust. J. agric. Res.* **16**, 825–36.

CARTER, M. V., and BANYER, R. J. (1964) Periodicity of basidiospore release in *Puccinia malvacearum. Aust. J. biol. Sci.* **17**, 801–2.

CARTWRIGHT, K. ST. G. (1926) Notes on a fungus associated with *Sirex cyaneus. Ann. appl. Biol.* **16**, 184–7.

CHADEFAUD, M. (1969) Remarques sur les parois, l'appareil apical et les reserves nutritive des asques. *Osterr. Bot. Z.* **116**, 181–202.

CHAMBERLAIN, A. C. (1967) Deposition of particles to natural surfaces. In *Airborne microbes, Proceeding of the seventeenth symposium of the Society for General Microbiology* (Editors P. H. Gregory, and J. L. Monteith), pp. 138–64. Cambridge.

CORBIN, J. B., OGAWA, J. M., and SCHULTZ, H. B. (1968) Fluctuations in numbers of *Monilinia laxa* conidia in an apricot orchard during the 1966 season. *Phytopathology* **58**, 1387–94.

CORKE, A. T. K. (1966) The role of rainwater in the movement of *Gloeosporium* spores on apple trees. In *The fungus spore, Proceedings of the eighteenth symposium of the Colston Research Society, Bristol,* 1966 (Editor M. F. Madelin), pp. 143–9. Butterworth's, London.

CORNER, E. J. H. (1929) Studies in the morphology of Discomycetes. II. The structure and development of the ascocarp. *Trans. Br. mycol. Soc.* **14**, 275–91.

—— (1948) Studies in the basidium I. The ampoule effect, with a note on nomenclature. *New Phytol.* **47**, 22–51.

—— (1950) *A monograph of Clavaria and allied genera.* London.

COUCH, J. N. (1941) The structure and action of cilia in some aquatic Phycomycetes. *Am. J. Bot.* **28**, 704–13.

CRAIGIE, J. H. (1931) An experimental investigation of sex in the rust fungi. *Phytopathology* **21**, 1001–40.

—— (1945) Epidemiology of stem rust in Western Canada. *Scient. Agri.* **25**, 285–401.

CUNNINGHAM, D. D. (*c.* 1873) *Microscopic examination of air.* Calcutta.

DAVIES, R. R. (1957) A study of air-borne *Cladosporium. Trans. Br. mycol. Soc.* **40**, 409–14.

—— (1959) Detachment of conidia by cloud droplets. *Nature, Lond.* **183**, 1695.

—— (1961) Wettability and capture, carriage and deposition of particles by raindrops. *Nature, Lond.* **191**, 616–17.

DE BARY, A. (1887) *Fungi, Mycetozoa and Bacteria* (English trans.). Oxford.

DERX, H. G. (1948) *Itersonilia,* nouveau genre de Sporobolomycètes à mycélium bouclé. *Bull. bot. Gdns Buitenz.* Ser. III, **18**, 465–82.

DIXON, H. H. (1914) *Transpiration and the ascent of sap in plants.* London.

DIXON, P. A. (1963) Spore liberation by water drops in some Myxomycetes. *Trans. Br. mycol. Soc.* **46**, 615–19.

—— (1965) The development and liberation of the conidia of *Xylosphaera furcata. Trans. Br. mycol. Soc.* **48**, 211–17.

DOBBS, C. G. (1939). Spore dispersal in the Mucorales. *Nature, Lond.* **149**, 583.

DODGE, B. O. (1924). Aecidiospore discharge as related to the character of the spore wall. *J. agric. Res.* **27**, 749–56.

DOGUET, G. (1962*a*) *Digitatispora marina* n.g., n.sp., basidiomycète marin. *C.r. hebd. Séanc. Acad. Sci., Paris* **254**, 4336–8.

DOGUET, G. (1962*b*) Récherches sur les noyaux des basides du *Digitatispora marina. Bull. Soc. mycol. Fr.* 78, 283–90.

—— (1967) *Nia vibrissa* Moore et Meyers, remarquable basidiomycète marin. *C.r. hebd. Séanc. Acad. Sci., Paris,* Sér. D., 265, 1780–3.

DOWDING, E. S. (1955) *Endogone* in Canadian rodents. *Mycologia* 47, 51–7.

DOWDING, P. (1969) The dispersal and survival of spores of fungi causing bluestain in pine. *Trans. Br. mycol. Soc.* 52, 125–37.

DOWSON, W. J. (1925) A die-back of rambler roses due to *Gnomonia rubi* Rehm. *J. R. hort. Soc.* 50, 55–72.

DUKES, P. D., and APPLE, J. L. (1961) Chemotaxis of zoospores of *Phytophthora parasitica* var. *nicotianae* by plant roots and certain chemical solutions. *Phytopathology* 51, 195–7.

DURHAM, O. C. (1942) Air-borne fungus spores as allergens. In *Aerobiology.* Washington.

ENGEL, H., and FRIEDERICHSEN, I. (1964) Der Abschuss der Sporangiolen von *Sphaerobolous stellatus* (Thode) Pers. in kintinuierlicher Dunkelheit. *Planta* 61, 361–70.

—— and SCHNEIDER, J. C. (1963) Die Umwandlung von Glycogen in Zucker in den Fruchtkorpern von *Sphaerobolus stellatus* (Thode) Pers. vor ihrem Abschusz. *Ber. dt. bot. Ges.* 75, 397–400.

ESSER, K., and STRAUB, J. (1958) Genetische Untersuchungen an *Sordaria macrospora* Auersw., Kompensation und Induktion bei genbedingten Entwicklungsdefeckten. *Z. VererbLehre* 80, 729–46.

FALCK, R. (1948) *Grundlinien eines orbis-vitalen Systems der Fadenpilze.* Amsterdam.

FAULWETTER, R. C. (1917*a*) Dissemination of the angular leafspot of cotton. *J. agric. Res.* 8, 457–75.

—— (1917*b*) Wind-blown rain, a factor in disease dissemination. *J. agric. Res.* 10, 639–48.

FISCHER, F. G. and WERNER, G. (1958) Die Chemotaxis der Schwarmsporen von Wasserpilzen (Saprolegniaceen). *Hoppe-Seyler's Z. physiol. Chem.* 310, 65–91.

FORESTRY COMMISSION (1947) Elm disease (*Ceratostomella ulmi*) *Leafl. For. Comm.* 19, 1–8.

FRASER, H. L. (1944) Observations on the method of transmission of internal boll disease of cotton by the cotton stainer-bug. *Ann. appl. Biol.* 31, 271–90.

FREEMAN, E. M. (1904) The seed fungus of *Lolium temulentum* L. *Phil. Trans. R. Soc.* B 196, 1–27.

FRIEDERICHSEN, I., and ENGEL, H. (1960) Der Abschussrhythmus der Fruchtkopter von *Sphaerobolus stellatus* (Thode) Pers. *Planta* 55, 313–26.

GARETH JONES, E. B. (1964) *Nautosphaeria cristaminuta* gen. et sp. nov., a marine pyrenomycete on submerged wood. *Trans. Br. mycol. Soc.* 47, 97–101.

GAY, J. L., and GREENWOOD, A. D. (1966) Structural aspects of zoospore production in *Saprolegnia ferax* with particular reference to the cell and vacuolar membranes. In *The fungus spore, Proceedings of the eighteenth symposium of the Colston Research Society, Bristol,* 1966 (Editor M. F. Madelin), pp. 95–108. Butterworths, London.

GEIGER, R. G. (1965) *The climate near the ground.* Cambridge, Mass.

GLYNNE, M. D. (1953) Production of spores by *Cercosporella herotrichoides*. *Trans. Br. mycol. Soc.* **36**, 46–51.

GREGORY, P. H. (1945) The dispersion of air-borne spores. *Trans. Br. mycol. Soc.* **28**, 26–72.

—— (1951) Deposition of air-borne *Lycopodium* spores on cylinders. *Ann. appl. Biol.* **38**, 357–76.

—— (1952) Fungus spores. *Trans. Br. mycol. Soc.* **35**, 1–18.

—— (1957) Electrostatic charges on spores of fungi in air. *Nature, Lond.* **180**, 330.

—— (1961) *The microbiology of the atmosphere*. London.

—— GUTHRIE, E. J., and BUNCE, E. (1959) Experiments on splash dispersal of fungus spores. *J. gen. Microbiol.* **20**, 328–54.

—— and LACEY, M. E. (1963) Liberation of spores from mouldy hay. *Trans. Br. mycol. Soc.* **46**, 73–80.

—— —— (1964) The discovery of *Pithomyces chartarum* in Britain. *Trans. Br. mycol. Soc.* **47**, 25–30.

—— and SREERAMULU, T. (1958) Air spora of an estuary. *Trans. Br. mycol. Soc.* **41**, 145–56.

—— and STEDMAN, O. J. (1953) Deposition of air-borne *Lycopodium* spores on plane surfaces. *Ann. appl. Biol.* **40**, 651–74.

—— —— (1958) Spore dispersal in *Ophiobolus graminis* and other fungi of cereal foot rots. *Trans. Br. mycol. Soc.* **41**, 449–56.

—— and WALLER, S. (1951) *Cryptostroma corticale* and sooty bark disease of sycamore (*Acer pseudoplatanus*). *Trans. Br. mycol. Soc.* **34**, 579–97.

GRUENHAGEN, R. E. (1945) *Hypoxylon pruinatum* and its pathogenesis on poplar. *Phytopathology* **35**, 72–89.

GRÜSZ, J. (1927) Genetische und gärungsphysiologische Untersuchungen an Nektarhefen. *Jb. wiss. Bot.* **66**, 109–82.

HADLEY, G. (1968) Development of stromata in *Claviceps purpurea*. *Trans. Br. mycol. Soc.* **51**, 763–9.

HAFIZ, A. (1951) Cultural studies of *Ascochyta rabiei* with special reference to zonation. *Trans. Br. mycol. Soc.* **34**, 259–69.

HALL, M. P. (1933) An analysis of the factors controlling growth form of certain fungi, with special reference to *Sclerotinia* (*Monilia*) *fructigena*. *Ann. Bot.* **47**, 543–78.

HARLEY, J. L. (1950) Recent progress in the study of endotrophic mycorrhiza. *New Phytol.* **49**, 213–47.

HARPER, J. E., and WEBSTER, J. (1964) An experimental analysis of the copro-philous fungus succession. *Trans. Br. mycol. Soc.* **47**, 511–30.

HARVEY, R. (1967) Air-spora studies at Cardiff I. *Cladosporium*. *Trans. Br. mycol. Soc.* **50**, 479–95.

—— HODGKISS, I. J., and LEWIS, P. N. (1969) Air-spora studies at Cardiff. II. *Chaetomium*. *Trans. Br. mycol. Soc.* **53**, 269–78.

HAUTMANN, F. (1924) Über die Nektarhefe *Anthomyces Reukaufii*. *Arch. Protistenk.* **48**, 213–44.

HEALD, F. H., and STUDHALTER, R. A. (1914) Birds as carriers of the chestnut blight fungus. *J. agric. Res.* **2**, 405–22.

HEIM, R. (1948) Phylogeny and natural classification of macro-fungi. *Trans. Br. mycol. Soc.* **30**, 161–78.

HIRST, J. M. (1952) An automatic volumetric spore trap. *Ann. appl. Biol.* 39, 257–65.

—— (1953) Changes in atmospheric spore content: diurnal periodicity and the effects of weather. *Trans. Br. mycol. Soc.* 36, 375–93.

—— and STEDMAN, O. J. (1961) The epidemiology of apple scab (*Venturia inaequalis* (Cke) Wint.). I. Frequency of airborne spores in orchards. *Ann. appl. Biol.* 49, 290–305.

—— —— (1962) The epidemiology of apply scab (*Venturia inaequalis* (Cke) Wint.). III. The supply of ascospores. *Ann. appl. Biol.* 50, 551–67.

—— —— (1963) Dry liberation of fungus spores by raindrops. *J. gen. Microbiol.* 33, 335–44.

—— —— and HOGG, W. H. (1967) Long-distance spore transport: methods of measurement, vertical spore profiles and the detection of immigrant spores. *J. gen. Microbiol.* 48, 329–55.

—— —— and HURST, G. W. (1967) Long-distance spore transport: vertical sections of spore clouds over the sea. *J. gen. Microbiol.* 48, 357–77.

—— STOREY, I. F., WARD, W. C., and WILCOX, H. J. (1955) The origin of apple scab epidemics in the Wisbech area in 1953 and 1954. *Pl. Path.* 4, 91–6.

HODGETTS, W. J. (1917) On the forcible discharge of spores of *Leptosphaeria acuta*. *New Phytol.* 16, 139–46.

HODGKISS, I. J., and HARVEY, R. (1969) Spore discharge rhythms in Pyrenomycetes. VI. The effects of climatic factors on seasonal and diurnal periodicities. *Trans. Br. mycol. Soc.* 52, 355–63.

HOLLIDAY, P. (1969) Dispersal of conidia of *Dothidella ulei* from *Hevea brasiliensis*. *Ann. appl. Biol.* 63, 435–47.

HUBBARD, H. G. (1892) The inhabitants of a fungus. *Can. Ent.* 24, 250–6.

HYDE, H. A., and WILLIAMS, D. A. (1946) A daily census of *Alternaria* spores caught from the atmosphere at Cardiff in 1942 and 1943. *Trans. Br. mycol. Soc.* 29, 78–85.

INGOLD, C. T. (1933) Spore discharge in the Ascomycetes. *New Phyol.* 32, 178–96.

—— (1934) The spore discharge mechanism of *Basidiobolus ranarum*. *New Phytol.* 33, 274–7.

—— (1939) *Spore discharge in land plants.* Oxford.

—— (1942) Aquatic Hyphomycetes of decaying alder leaves. *Trans. Br. mycol. Soc.* 25, 339–417.

—— (1946*a*) Size and form in agarics. *Trans. Br. mycol. Soc.* 29, 108–13.

—— (1946*b*) Spore discharge in *Daldinia concentrica*. *Trans. Br. mycol. Soc.* 29, 43–51.

—— (1948) The water-relations of spore discharge in *Epichloe*. *Trans. Br. mycol. Soc.* 31, 277–80.

—— (1954*a*) Ascospore form. *Trans. Br. mycol. Soc.* 37, 19–21.

—— (1954*b*) Fungi and water. *Trans. Br. mycol. Soc.* 37, 97–107.

—— (1956*a*) A gas phase in viable fungal spores. *Nature, Lond.* 177, 1242–3.

—— (1956*b*) The spore deposit of *Daldinia Trans. Br. mycol. Soc.* 39, 378–80.

—— (1957) Spore liberation in higher fungi. *Endeavour* 16, 78–83.

—— (1958) On light-stimulated spore discharge in *Sordaria*. *Ann. Bot.* 22, 129–35.

—— (1959) Jelly as a water-reserve in fungi. *Trans. Br. mycol. Soc.* **42**, 475–8.

—— (1960) Spore discharge in Pyrenomycetes. *Friesia* **6**, 148–63.

—— (1961) Ballistics of certain Ascomycetes. *New Phytol.* **60**, 143–9.

—— (1964) Possible spore discharge mechanism in *Pyricularia*. *Trans. Br. mycol. Soc.* **47**, 573–5.

—— (1965) *Spore liberation*. Oxford.

—— (1966) Aspects of spore liberation: violent discharge. In *The fungus spore, Proceedings of the eighteenth symposium of the Colston Research Society, Bristol,* 1966 (Editor M. F. Madelin), pp. 113–32. Butterworths, London.

—— (1966) The tetraradiate aquatic fungal spore. *Mycologia* **58**, 43–6.

—— (1968) Spore liberation in *Loramyces*. *Trans. Br. mycol. Soc.* **51**, 323–5.

—— (1969) The ballistospore. *Friesia* **9**, 66–76.

—— and Cox, V. J. (1955) Periodicity of spore discharge in *Daldinia*. *Ann. Bot.* **29**, 201–9.

—— and Dann, V. (1968) Spore discharge in fungi under very high surrounding air-pressure, and the bubble-theory of ballistospore release. *Mycologia* **60**, 285–9.

—— —— (1968) Spore-size and discharge in *Conidiobolus*. *Trans. Br. mycol. Soc.* **51**, 588–91.

—— and Dring, V. J. (1957) An analysis of spore discharge in *Sordaria*. *Ann. Bot.* **21**, 465–77.

—— and Hadland, S. A. (1959) The ballistics of *Sordaria*. *New Phytol.* **58**, 46–57.

—— and Marshall, B. (1962) Stimulation of spore discharge by reduced humidity in *Sordaria*. *Ann. Bot.* **26**, 563–8.

—— —— (1963) Further observations on light and spore discharge in certain Pyrenomycetes. *Ann. Bot.* **27**, 481–91.

—— —— (1964) Stimulation of spore discharge in *Sordaria* by carbon dioxide. *Ann. Bot.* **28**, 325–9.

—— and Nawaz, M. (1967) Sporophore production in *Sphaerobolus* with special reference to periodicity. *Ann. Bot.* **31**, 791–802.

—— and Oso, B. A. (1968) Increased distance of discharge due to puffing in *Ascobolus*. *Trans. Br. mycol. Soc.* **51**, 592–4.

—— (1969) Light and spore discharge in *Ascobolus*. *Ann. Bot.* **33**, 463–71.

—— and Zoberi, M. H. (1963) The asexual apparatus of Mucorales in relation to spore liberation. *Trans. Br. mycol. Soc.* **46**, 115–34.

Jarvis, W. R. (1962) The dispersal of spores of *Botrytis cinerea* Fr. in a raspberry plantation. *Trans. Br. mycol. Soc.* **45**, 549–59.

Kohlmeyer, J. (1960) Wood-inhabiting marine fungi from the Pacific Northwest and California. *Nova Hedwigia* **2**, 293–343.

—— (1962) The importance of fungi in the sea. Proceedings of the *Symposium on marine microbiology* (Editor C. H. Oppenheimer), pp. 300–314. Springfield, Illinois.

—— (1963) Fungi marini novi vel critici. *Nova Hedwigia* **6**, 297–329.

Kramer, C. L., and Pady, S. M. (1966) A new 24-hour spore sampler. *Phytopathology* **56**, 517–20.

—— —— (1968a) Viability of airborne spores. *Mycologia* **60**, 448–9.

—— —— (1968b) Spore discharge in *Hypocrea gelatinosa*. *Mycologia* **60**, 208–10.

Kramer, C. L., Pady, S. M., Clary, R., and Haard, R. (1968) Diurnal periodicity in aeciospore release of certain rusts. *Trans. Br. mycol. Soc.* 51, 679–87.
—— —— Rogerson, C. T., and Ouye, L. G. (1959) Kansas Aeromycology II. Materials, methods and general results. *Trans. Kans. Acad. Sci.* 62, 184–99.
Lacey, M. E. (1962) The summer air-spora of two contrasting adjacent rural sites. *J. gen. Microbiol.* 29, 485–501.
Large, E. C. (1961) Pursuits of mycology. *Trans. Br. mycol. Soc.* 44, 1–23.
Lange, J. E. (1946) *Flora Agaricina Danica.* Copenhagen.
Leach, J. G. (1940) *Insect transmission of plant diseases.* New York.
—— Clinton, St. J. P. and Christensen, C. M. (1940) Observations on two ambrosia beetles and their associated fungi. *Phytopathology* 30, 227–36.
Levisohn, I. (1927) Beitrag zur Entwicklungsgeschichte und Biologie von *Basidiobolus ranarum* Eidam. *Jb. wiss. Bot.* 66, 513–55.
Lortie, M. and Kuntz, J. E. (1963) Ascospore discharge and conidium release by *Nectria galligena* Bres. under field and laboratory conditions. *Can. J. Bot.* 41, 1203–10.
McKeen, W. E. (1962) The flagellation, movement, and encystment of some phycomycetous zoospores. *Can. J. Microbiol.* 8, 897–904.
McOnie, K. C. (1964) Orchard development and discharge of ascospores of *Guignardia citricarpa* and the onset of infection in relation to the control of citrus black spot. *Phytopathology* 54, 1448–53.
Machlis, L. (1968) Zoospore chemotaxis in the watermold *Allomyces*. *Physiologia Pl.* 22, 126–39.
Malone, J. P., and Muskett, A. E. (1964) Seed-borne fungi. *Proc. int. Seed Test. Ass.* 29, 179–384.
Martin, G. W. (1925) Morphology of *Conidiobolus villosus*. *Bot. Gaz.* 80, 311–18.
Martinson, F. D., and Clark, B. M. (1967) Rhinophycomycosis entomophthorae in *Nigeria*. *Am. J. trop. Med. Hyg.* 16, 40–7.
May, K. R. (1945) The cascade impactor: an instrument for sampling coarse aerosols. *J. scient. Instrum.* 22, 187–95.
—— (1967) Physical aspects of sampling airborne microbes. In *Airborne microbes, Proceedings of the seventeenth symposium of the Society for General Microbiology* (Editors P. H. Gregory and J. L. Monteith), pp. 60–80. Cambridge.
Meredith, D. S. (1961) Spore discharge in *Deightoniella torulosa* (Syd.) Ellis. *Ann. Bot.* 25, 271–8.
—— (1962a) Spore discharge in *Cordana musae* (Zimm.) Höhnel and *Zygosporium oscheoides* Mont. *Ann. Bot.* 26, 233–41.
—— (1962b) Dispersal of conidia of *Cordana musae* (Zimm.) Höhnel in Jamaican banana plantations. *Ann. appl. Biol.* 50, 263–7.
—— (1963) Further observations on the zonate eyespot fungus, *Drechslera gigantea*, in Jamaica. *Trans. Br. mycol. Soc.* 46, 201–7.
—— (1963) Violent spore release in some Fungi Imperfecti. *Ann. Bot.* 27, 39–47.
—— (1965) Violent spore release in *Helminthosporium turcicum*. *Phytopathology* 55, 1099–02.
Miquel, P. (1883) *Les organismes vivantes de l'atmosphere.* Paris.
Moller, W. J., and Carter, M. V. (1965) Production and dispersal of ascospores in *Eutypa armeniacae*. *Aust. J. biol. Sci.* 18, 67–80.

Müller, D. (1954) Die Abschleudeuderung der Sporen von *Sporobolomyces*—Spiegelhefe—gefilmt. *Friesia* **5**, 65–74.

Moore, R. T. (1966) Basidiospore discharge: a generalized concept based on cell biology. *Science, N.Y.* **154**, 424.

Muskett, A. E. (1950) Seed-borne fungi and their significance. *Trans. Br. mycol. Soc.* **33**, 1–12.

—— and Colhoun, J. (1947) *The diseases of the flax plant.* Belfast.

Nawaz, M. (1967) Phototropism in *Sphaerobolus. Biologia* **13**, 5–14.

Olive, L. S. (1964) Spore discharge mechanism in Basidiomycetes. *Science, N.Y.* **146**, 542–543.

—— and Stoianovitch, C. (1966) A simple new mycetozoan with ballistospores. *Am. J. Bot.* **53**, 344–9.

Oso, B. A. (1969*a*) Effects of temperature and mechanical agitation on spore discharge in *Ascobolus crenulatus. Trans. Br. mycol. Soc.* **53**, 291–7.

—— (1969*b*) Spore discharge in Discomycetes in relation to relative humidity. *Trans. Br. mycol. Soc.* **53**, 119–26.

Pady, S. M., and Gregory, P. H. (1963) Numbers and viability of airborne hyphal fragments in England. *Trans. Br. mycol. Soc.* **46**, 609–13.

—— and Kapica, L. (1955) Fungi in air over the Atlantic Ocean. *Mycologia* **47**, 34–50.

—— and Kramer, C. L. (1969) Periodicity in ascospore discharge in *Bombardia. Trans. Br. mycol. Soc.* **53**, 449–54.

—— Kramer, C. L., and Clary, R. (1967) Diurnal periodicity in airborne fungi in an orchard. *J. Allergy* **39**, 302–10.

—— —— —— (1968) Periodicity in aeciospore release in *Gymnosporangium juniperi-virginianae. Phytopathology* **58**, 329–31.

—— —— —— (1969*a*) Sporulation in some species of *Erysiphe. Phytopathology* **59**, 844–8.

—— —— —— (1969*b*) Aeciospore release in *Gymnosporangium. Can. J. Bot.* **47**, 1027–32.

—— —— —— (1969*c*) Periodicity in spore release in *Cladosporium. Mycologia* **61**, 87–98.

Page, R. M. (1962) Light and asexual reproduction of *Pilobolus. Science, N.Y.* **138**, 1238–45.

—— (1964) Sporangium discharge in *Pilobolus:* a photographic study. *Science, N.Y.* **146**, 925–7.

—— and Kennedy, D. (1964) Studies on the velocity of discharged sporangia of *Pilobolus kleinii. Mycologia* **56**, 363–8.

Parkin, E. A. (1942) Symbiosis and siricid wood-wasps. *Ann. appl. Biol.* **29**, 268–74.

Pasteur, L. (1862) Mémoire sur les corpuscules organisés qui existent dans l'atmosphère. *Annls Chim. Phys.* **64**, 5–110.

Pathak, V. K., and Pady, S. M. (1965) Numbers and viability of certain airborne fungus spores. *Mycologia* **57**, 301–10.

Pegler, D. N., and Young, T. W. K. (1969) Ultrastructure of basidiospores in Agaricales in relation to taxonomy and spore discharge. *Trans. Br. mycol. Soc.* **52**, 491–6.

PFEFFER, W. (1884) Lokmotorische Richtungsbewegungen durch chemische Reize. *Unterr. Bl Bot. Inst. Tübingen* 1, 363–481.

PINCKARD, J. A. (1942) The mechanism of spore dispersal in *Peronospora tabacina* and certain other downy mildew fungi. *Phytopathology* 32, 505–11.

PLUNKETT, B. E. (1961) The change of tropism in *Polyporus brumalis* stipes and the effect of directional stimuli on pileus differentiation. *Ann. Bot.* 25, 206–23.

PRINCE, A. E. (1943) Basidium formation and spore discharge in *Gymnosporangium nidus-avis*. *Farlowia* 1, 79–93.

PRINGSHEIM, E. G., and CZURDA, V. (1927) Phototropische und ballistische Probleme bei *Pilobolus*. *Jb. wiss. Bot.* 66, 863–901.

PRINGSHEIM, N. (1858) Ueber das Austreten der Sporen von *Sphaeria Scirpi* aus ihren Schlauchen. *Jb. wiss. Bot.* 1, 189–92.

RAMSBOTTOM, J. (1923) *A handbook of the larger British fungi*. London.

—— (1941) The expanding knowledge of mycology since *Linnaeus*. *Proc. Linn. Soc. Lond.* 151, 280–367.

RÁTHAY, M. C. (1883) Untersuchungen über die Spermagonien der Rostpilze. *Denkscr. Akad. Wiss.*, *Wien* 46, 1–51.

ROMAGNESI, H. (1933) Le genre *Richoniella*, chaînon angiocarpe de la série des Rhodogoniosporés. *Bull. Soc. mycol. Fr.* 49, 433–4.

ROYLE, D. J., and HICKMAN, C. J. (1964) Analysis of factor governing *in vitro* accumulation of zoospores of *Pythium aphanidermatum* on roots. I. Behaviour of zoospores. *Can. J. Microbiol.* 10, 151–62.

—— —— (1964) Analysis of factors governing *in vitro* accumulation of zoospores of *Pythium aphanidermatum* on roots. II. Substances causing response. *Can. J. Microbiol.* 10, 201–19.

SAGROMSKY, H. (1952a) Der Einflusz des Liches auf die rhythmische Konidienbildung von *Penicillium*. *Flora, Jena* 139, 300–13.

—— (1952b) Lichtinduzierte Ringbildung bei Pilzen. II. *Flora, Jena* 139, 560–4.

SALMON, E. S. (1914) Observations on the perithecial stage of American Gooseberry Mildew (*Sphaerotheca mors-uvae* (Schwein.) Berk.). *J. agric. Sci.* 6, 187–93.

SAMPSON, K. (1933) The systemic infection of grasses by *Epichloe typhina* (Pers.) Tul. *Trans. Br. mycol. Soc.* 17, 30–49.

—— (1935) The presence and absence of an endophytic fungus in *Lolium temulentum* and *L. perenne*. *Trans. Br. mycol. Soc.* 19, 337–43.

—— (1937) Further observations on the systemic infection of *Lolium*. *Trans. Br. mycol. Soc.* 21, 84–97.

—— (1939) Additional notes on the systemic infection of *Lolium*. *Trans. Br. mycol. Soc.* 23, 316–19.

SAVILE, D. B. O. (1965) Spore discharge in Basidiomycetes: a unified theory. *Science, N.Y.* 147, 165–6.

SCHMIDLE, A. (1951) Die Tagesperiodizitat der asexuellen Reproduktion von *Pilobolus sphaerosporus*. *Arch. Mikrobiol.* 16, 80–100.

SCHRÖDTER, H. (1960) Dispersal by air and water—the flight and landing. In *Plant pathology: an advanced treatise*, Vol. 3, pp. 169–227, New York.

SILOW, R. A. (1933) A systemic disease of red clover caused by *Botrytis anthophila* Bond. *Trans. Br. mycol. Soc.* 18, 239–48.

SINGER, R., and SMITH, A. H. (1960) Studies on secotiaceous fungi XI. The astrogastraceous series. *Mem. Torrey bot. Club* 21, 1–112.

SMITH, R. S. (1966) The liberation of cereal stem rust uredospores under various environmental conditions in a wind tunnel. *Trans. Br. mycol. Soc.* 49, 33–41.

SNOW, G. A., and FROELICH, R. C. (1968) Daily and seasonal dispersal of basidiospores of *Cronartium fusiforme. Phytopathology* 58, 1532–6.

SREERAMULA, T. (1959) The diurnal and seasonal periodicity of spores of certain plant pathogens in the air. *Trans. Br. mycol. Soc.* 42, 177–84.

—— (1962) Aerial dissemination of barley loose smut (*Ustilago nuda*). *Trans. Br. mycol. Soc.* 45, 373–84.

—— (1964) Incidence of conidia of *Erysiphe graminis* in the air above a mildew-infected barley field. *Trans. Br. mycol. Soc.* 47, 31–8.

—— and RAMALINGAM, A. (1961) Experiments on the dispersion of *Lycopodium* and *Podaxis* spores in the air. *Ann. appl. Biol.* 49, 659–70.

—— —— (1964) Some short-period changes in the atmospheric spore content associated with changes in the weather and other conditions. *Proc. Indian Acad. Sci.* 59, 154–72.

—— and SESHAVATARAM, V. (1962) Spore content of air over paddy fields. I. Changes in a field near Pentapadu from 21 September to 31 December 1957. *Indian Phytopath.* 15, 61–74.

STAKMAN, E. C. (1947) International problems in plant disease control. *Proc. Am. phil. Soc.* 91, 95–111.

—— and CHRISTENSEN, C. M. (1946) Aerobiology in relation to plant disease. *Bot. Rev.* 12, 205–53.

—— HENRY, A. W., CURRAN, G. C., and CHRISTOPHER, W. N. (1923) Spores in the upper air. *J. agric. Res.* 24, 599–605.

STEPANOV, K. M. (1935) Translated title: Dissemination of infectious diseases of plants by air currents. *Trudy Zasch. Rast.* 2nd Ser. (Phytopathology) 8, 1–68.

SUSSMAN, A. S., and HALVORSON, H. O. (1966) *Spores their dormancy and germination.* New York.

SWINBANK, P., TAGGART, J., and HUTCHINSON, S. A. (1964) The measurement of electrostatic charges on spores of *Merulius lacrymans* (Wulf.) Fr. *Ann. Bot.* 28, 239–49.

TAGGART, J., HUTCHINSON, S. A., and SWINBANK, P. (1964) Spore discharge in Basidiomycetes. III. Spore liberation from narrow hymenial tubes. *Ann. Bot.* 28, 607–18.

THIRUMALACHAR, M. J., KERN, F. D., and PATIL, B. V. (1966) *Elateraecium,* a new form genus of Uredinales. *Mycologia* 58, 391–6.

TURNER, D. M., and MILLARD, W. A. (1931) Leaf-spot of oats, *Helminthosporium avenae* (Bri. and Cav.) Eid. *Ann. appl. Biol.* 18, 535–58.

UEBELMESSER, E. R. (1954) Über den endonomen Tagesrhythmus der Sporangientragerbildung von *Pilobolus. Arch. Mikrobiol.* 20, 1–33.

UPHOF, J. C. T. (1942) Ecological relations of plants with ants and termites. *Bot. Rev.* 8, 563–98.

VAN ARSDEL, E. P. (1965) Micrometeorology and plant disease epidemiology. *Phytopathology* 55, 945–50.

VAN DER PLANK, J. E. (1963) *Plant diseases: epidemics and control.* New York.

VOGLINO, P. (1895) Richerche intorno all' azione della lumacha e dei rospi nello svilappo di Agaricini. *Nuovo G. bot. ital.* **27**, 181–5.

WAGGONER, P. E., and TAYLOR, G. S. (1958) Dissemination by atmospheric turbulence: spores of *Peronospora tabacina*. *Phytopathology* **48**, 46–51.

WALKER, L. B., and ANDERSEN, E. N. (1925) Relation of glycogen to spore-ejection. *Mycologia* **15**, 154–9.

WALKEY, D. G. A., and HARVEY, R. (1966*a*) Studies of the ballistics of asco-spores. *New Phytol.* **65**, 59–74.

—— —— (1966*b*) Spore discharge rhythms in pyrenomycetes. I. A survey of the periodicity of spore discharge in pyrenomycetes. *Trans. Br. mycol. Soc.* **49**, 583–92.

WEBB, S. (1945) Australian ambrosia fungi. *Proc. R. Soc. Vict.* **57**, 57–78.

WEBBER, F. C. (1967) Observations on the structure, life history and biology of *Mycosphaerella ascophylli*. *Trans. Br. mycol. Soc.* **50**, 583–601.

WEBSTER, J. (1952) Spore discharge in the hyphomycete *Nigrospora sphaerica*. *New Phytol.* **51**, 229–35.

—— (1959) Experiments with spores of aquatic hyphomycetes. I. Sedimentation and impaction on smooth surfaces. *Ann. Bot.* **23**, 595–611.

—— (1966) Spore projection in *Epicoccum* and *Arthrinium*. *Trans. Br. mycol. Soc.* **49**, 339–43.

—— and DENNIS, C. (1967) The mechanism of sporangial discharge in *Pythium middletonii*. *New Phytol.* **66**, 307–13.

WEIMER, J. L. (1917) Three cedar rust fungi. Their life histories and the diseases they produce. *Cornell University Agricultural Experimental Station Bulletin No.* 390, pp. 509–49.

WELLS, K. (1965) Ultrastructural features of developing and mature basidia and basidiospores of *Schizophyllum commune*. *Mycologia* **57**, 236–61.

WESTON, W. H. (1923) Production and dispersal of conidia in the Philippine sclerosporas of maize. *J. agric. Res.* **23**, 239–78.

WILLOUGHBY, L. G. (1962) The occurrence and distribution of reproductive spores of Saprolegniales in fresh water. *J. Ecol.* **50**, 733–59.

WILSON, M., NOBLE, M., and GRAY, E. G. (1945) The blind-seed disease of rye-grass and its causal fungus. *Trans. R. Soc. Edinb.* **61**, 327–340.

WOODWARD, R. C. (1927) Studies on *Podosphaera leucotricha* (Ell. and Ev.) Salm. *Trans. Br. mycol. Soc.* **12**, 173–204.

YARWOOD, C. E. (1936) The diurnal cycle of the powdery mildew *Erysiphe polygoni*. *J. agric. Res.* **52**, 645–57.

ZADOKS, J. C. (1965) Epidemiology of wheat rusts in Europe. *Pl. Prot. Bull. F.A.O.* **13**, 1–12.

ZALEWSKI, A. (1883) Über Sporenabschnürung und Sporenabfallen bei den Pilzen. *Flora, Jena* **66**, 268–70.

ZOBERI, M. H. (1961) Take-off of mould spores in relation to wind speed and humidity. *Ann. Bot.* **25**, 53–64.

—— (1964) Effect of temperature and humidity on ballistospore discharge. *Trans. Br. mycol. Soc.* **47**, 109–14.

INDEX